Introduction to Theoretical Biology

A Theory of Heat

Daniel S. Szumski, Independent Scholar

The Unusual F Street Winery Press
Davis, California

Thanks to Martin Barnes for his careful editing of the manuscript

Author's Preface

This research had its beginnings in an engineer's curious mind and an unanswered question in aquatic toxicology. The initial engineering question was: "What causes the pollutant, ammonia, to be toxic to fish?" This soon developed to "What is free energy?", then "How does a cell work?", and finally Schrodinger's question, "What is Life?" The twists and turns in this pathway were a learning experience in independent scholarship that has held my interest for 35 years.

The question that I am most frequently asked is "Why don't you publish your work?" I wish I could. But, it is not that easy to get theoretical work published without an academic affiliation and credentials. I have neither. The answer to that question that I find most satisfying as an independent scholar is " My work is so far from the existing paradigms that no one in accepted professional circles wants to be associated with it". This is not necessarily a shortcoming. In fact, most independent scholars will tell you about the freedom in not being tied to the dogma and the current paradigms in their field of study. However, the lesson that each of us has to learn, is the value in knowing those paradigms so our work incorporates their truth, even as we focus on avoiding their limitations.

This book contains four technical papers that I collectively call <u>Theory of Heat</u>. All of this work is based on a far-from-equilibrium blackbody formula that I had to develop to understand free energy storage in biological systems, and more specifically, in a living cell. The engineer and mathematical modeler in me knew that it was senseless to try to model the cell's molecular quantities. There are simply too many. It made more sense, particularly in light of the types of free energy questions that I was asking, that I base my model in the energy that 'glues' the atoms together, and in particular, its far-from-equilibrium structure. It was likely that this energy's underlying structure was as far-from-equilibrium as the molecular quantities that virtually everyone else was studying.

In Chapter 1, I present basic concepts that are fundamental to theoretical biology. The most important of these is negative entropy production in its relationship to equilibrium and far-from-equilibrium thermodynamic states. Biological modeling fundamentals are introduced, and discussed briefly.

Chapter 2 contains an abbreviated version of my original far-from-equilibrium blackbody theory. It is the basis for everything that is to follow because it is here that I introduce a new quantity to the study of physics and biophysics, the radiation temperature. It is identical to the thermodynamic temperature at equilibrium, but leads to profound conclusions when it is separated from the thermodynamic temperature in a far-from-equilibrium way. This chapter introduces thermodynamically reversible energy storage, and describes its place in relationship to irreversible process energy.

Chapter 3 presents the second of two papers describing a living cell model. It is presented in this order because it is the place where the far-from-equilibrium blackbody theory is first used. It asks questions about cell size and aging, about cancer…all from the standpoint of energy organization within the cell's covalent bond structure. This 'Theory of the Living State' describes the precision of cell synthesis. In it, one begins to see how a precise sequence of energy utilization steps evolved to regulate molecular events beyond DNA transcription. We begin to see how the gene is not isolated in the DNA, but rather in the overall temporal sequence of cell synthesis steps that has evolved over time.

The fourth chapter addresses some of the 'modeling framework' issues which one faces in building a cell model. It is a paper that I had developed in the early 1980's when my research was only beginning. This was my first foray into the world of reversible thermodynamics, and its consequences in cell model development. I was working here with a holistic model of trans-membrane ion transport that had come out of my efforts to model ammonia toxicity at a fish's gill epithelial surface. At the time, I had no idea that this would prove to be the deterministic methodology for explaining the variability in biological response. In the intervening years, I had learned how the Principle of Least action was operative in all reversible processes, and how it alone specified the next step in any thermodynamically reversible process. Here was a path to the variability of cellular metabolism in a wholly deterministic context. This was the critical insight that came out of my cold fusion research.

The final chapter is my cold fusion paper. It is a 'verification' of the Theory of Heat…its application to a second mysterious heat process. It does well as a predictive tool. This is the place where I learned to use the Principle of Least Action to determine, with incredible precision and exactitude, the next step in any thermodynamically reversible process.

Among the unwritten chapters is a theory of Sono-luminescence. It too uses the Theory of Heat's transference of energy between the mass and radiation domains, in this case to produce light from mechanical energy. I offer it as a challenge to my reader… to do some independent scholarship, and in the process, discover something new and delicious. This is my life's work. I hope that it gives you, the reader, the same level of satisfaction that it has always given me.

Dan Szumski, Independent Scholar
Davis, California, danszumski@gmail.com
September 6, 2013

Table of Contents

Chapter 1

What is Life?

A. Introduction

The process underlying matter's living state is probably very exact. All laws of nature are. But this one is particularly stubborn. It refuses to reveal itself in spite of the extraordinary expertise and experimental sophistication that have been brought to its scientific study. I submit, that it is likely that new science, and a radically different way of looking at other existing science, will be required before this inquiry is over.

A suitable theory of the living state will reveal the working of very exact natural laws. It will explain how these act to produce what Schrodinger called life's neg-entropic character. It will then go on to show how every step in cell synthesis and metabolic function, is precisely determined by the internal and external states that exist at that moment in the cell's life. Nature would not have it any other way. Reproductive fidelity and environmental response cannot be left to chance. They are molded by natural selection to produce exact outcomes, and in so doing; protect the organism to the extent possible.

'New Science' is Required

Our science is in its formative stages. We are still trying to understand the interconnectedness of cellular processes, while at the same time, we puzzle over the 'Big Picture'. We have not yet entered the golden age of biological science, where theory is stated in mathematics, and our calculation precision rivals that in physics. To achieve this plateau, we will have to understand at a very fundamental level, what it is that makes life go forward in time. This is the first goal of theoretical biology.

What we need now, more than anything else, is context. This seems to be well understood in the profession. The phrases 'quantitative biology' and 'theoretical biology' have become common in our lexicon. They envision a transformed science where we accept…no, require… the same types of mathematical formality that we see in mechanics, chemistry, and physics. But despite all of our efforts, there is still no clearly defined path to that objective; a circumstance that has persisted since Crick introduced his Central Dogma of molecular biology 64 years ago.

So what should we envision when we say that we are stepping back to look at the Big Picture? Well yes, it is that quest for context. But rather than trying to understand the common denominator in classes of biological processes, or formulating theoretical explanations for specific data sets, we should instead, be organizing everything that we know into a single theoretical framework that explains what we observe in our experiments. However, and here is the tricky part, in this process, we cannot introduce anything that is contrary to those observations. While most biologists believe that this can ultimately be done within the context of existing science, I must disagree.

It is my view, that we will first have to expand our understanding of free energy, probably with new science, and then be willing to entertain the possibility that reversible thermodynamic processes are operative at the most fundamental level in matter's living state.

You will notice an eye that reappears on illustrations used throughout this presentation. This is the eye of discovery. It indicates that there are new concepts here, and that your understanding will be enhanced if you put on your new eyes so that you might see the world in a radically different way.

"The only real voyage of discovery consists not in seeking new landscapes, but in having new eyes; in seeing the universe through the eyes of another, one hundred others- in seeing the hundred universes that each of them sees."

Marcel Proust - In Search of Lost Time

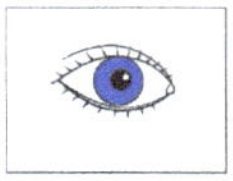

B. The Direction of Time's Arrow

Time is the progression of an entropic event. We experience it as the event sequence that always proceeds in one direction, and generally at a constant speed. It is an integral part of the theory that we hope to develop. However, our understanding of time needs to be clarified in two respects.

The first time measure that we want to explore is directional. Our positive time direction will be that in which the Second Law of Thermodynamics (the great law of entropy increase) prevails, and order is always and everywhere going over into disorder. This is the time direction wherein a weight falls according to Newtonian principles, celestial objects move and evolve, chemical reactions proceed to new equilibrium conditions, and (this is the important one in what is to follow) where dielectric loss continually changes three-dimensional electrical charge into the heat of three-dimensional, atomic and molecular motion.

But, more importantly, we will concern ourselves with the opposite time direction, a negative direction that when combined with our positive direction, modifies it, and slows the entropic time progression to where it perceptively slows, or even seems to stop. This is the time direction that dominates in living systems and gives them what Erwin Schrodinger described as life's neg-entropic character, that is, a series of events that proceeds to progressively more ordered states.

Secondly, we will want to explore the progression of events, using as a measure of time, its inverse...the frequency domain in which discrete events take place. We will use this as our measure of perception. It defines the time scale in which perceived events take place, and by extension, the sequence in which those events occur. This sequence can be either exact, as in the case of DNA transcription, or random, in a statistical-mechanical or thermodynamic sense. For example, randomness might describe in probabilistic terms, the transport of ions across a cell's membrane. Here too, we will be adding something entirely new, and this is important, ...a method for defining, from among all of the possible statistical mechanical events that can occur, precisely which one will occur next. We will return to this in Chapter 4.

But first, let's look at how time is relative depending upon the event that is occurring. Let's choose as an example, a chemical event, and in particular, the redox event: $2[H^+] + O_2 +$

$2\{e^-\} \overset{v_f}{\rightarrow} H_2O_2$, where the forward velocity of the reaction is $v_f = k_f[H^+]^2[O_2]\{e^-\}^2$, and k_f has the units per time, and v_f, the forward velocity of the reaction, is measured in $moles^5/time$.

However, this is only half of the overall reaction:

$$2[H^+] + [O_2] + 2\{e^-\} + e^{H_f'/k_bT} \overset{v_f}{\underset{v_b}{\longleftrightarrow}} [H_2O_2] + e^{H_f''/k_bT}$$

$$2[H^+] + [O_2] + 2\{e^-\} \overset{v_f}{\underset{v_b}{\longleftrightarrow}} [H_2O_2] + e^{\Delta H_f/k_bT}$$

where: $v_b = k_b[H_2O_2][\exp(\Delta H_f')]$

and $K_{eq}^o = {}^{v_f}/_{v_b} = \frac{k_b[H_2O_2][\exp(\Delta H'_f]}{k_f[H^+]^2[O_2]\{e^-\}^2}$, the true equilibrium constant for the reaction.

If we then set: $v_b = v_f$, $K_{eq}^o = 1.0$, and the apparent equilibrium constant, K_{eq}:

$$K_{eq} = {}^{k_f[e^{-\Delta H_f/k_bT}]}/_{k_b} = \frac{[H_2O_2]}{[H^+]^2[O_2]\{e^-\}^2}, \log K_{eq} = 23.1, E_H^o = 0.68\ volts$$

Forward, Backward and Net Velocity

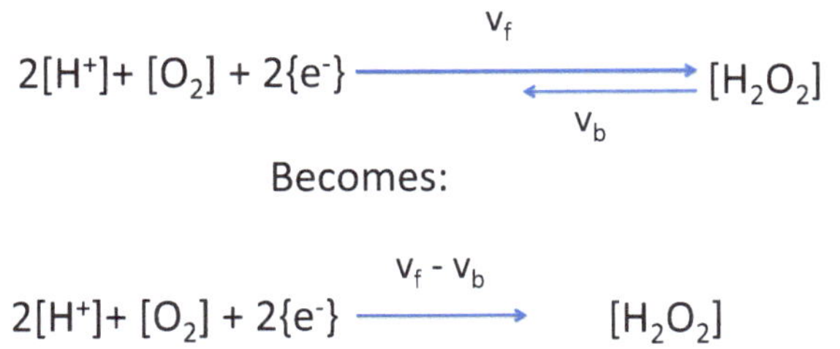

The positive value of $\log K_{eq}$ indicates a reaction that is heavily favored in the forward time direction. However, while the forward reaction may dominate, there is a small, but still significant reaction in what we would see as the negative time direction. In other words, we see a chemical event that results in H_2O_2 production in our positive time direction. But in looking more closely at the process, this is the net event, where a portion of its overall structure occurs in negative time, or the direction of entropy decrease. This reaction produces equilibrium states, and is purely entropic. We will call this the *normal entropic process*.

In an idealized world, it should be possible to alter the reaction to make the backward direction more dominant. This can be accomplished by altering either, or both velocities to make the difference between them smaller. This slows the overall reaction, and effectively decreases its entropy production. Schrodinger would say that we have made the process move in the negative entropy direction, or more simply, that it has become *negentropic* , a state in which the net reaction velocity is positive, but smaller than that of the *normal entropic process*. As the neg-entropic state develops, it moves farther-from-equilibrium, toward more and more improbable thermodynamic states, and increasing instability. From the time that the process left the normal entropic state it was a *negentropic process*.

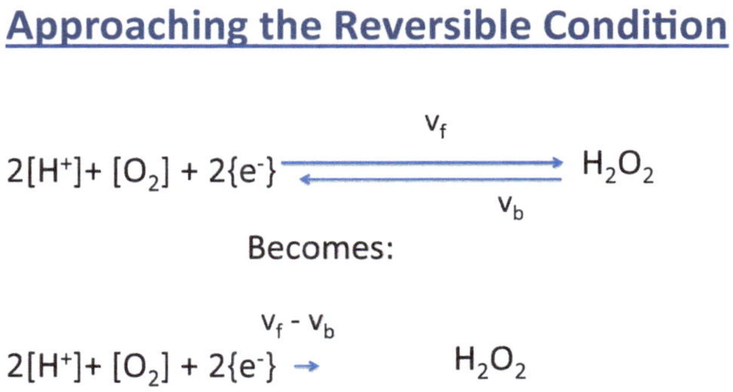

Now look what happens when we take this negentropic process to its limit; allowing the forward and backward reaction rates to become identical. This special case, places the process at the very limit of what the Second Law allows. Entropy production ceases, and the process is precisely balanced among all of its possible outcomes. There exists an equal probability of evolving in the forward and backward time dimensions, and all mass/energy and energy conversion outcomes are possible. Time appears to stand still. This is as far-from-equilibrium that the process being considered can go in the negentropic direction. We call this limiting case of the negentropic process, the *reversible process,* or the reaction's *reversible state*.

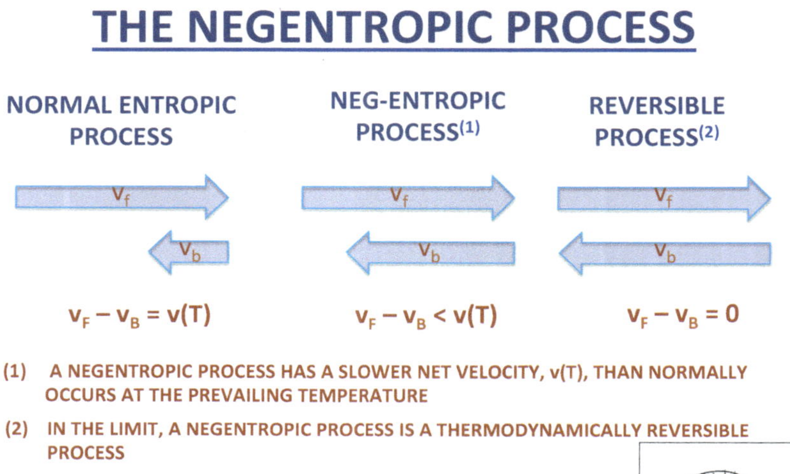

THE NEGENTROPIC PROCESS
NORMAL ENTROPIC PROCESS
NEG-ENTROPIC PROCESS(1)
REVERSIBLE PROCESS(2)
v_f
v_b
$v_F - v_B = v(T)$
$v_F - v_B < v(T)$
$v_F - v_B = 0$
(1) A NEGENTROPIC PROCESS HAS A SLOWER NET VELOCITY, v(T), THAN NORMALLY OCCURS AT THE PREVAILING TEMPERATURE
(2) IN THE LIMIT, A NEGENTROPIC PROCESS IS A THERMODYNAMICALLY REVERSIBLE PROCESS

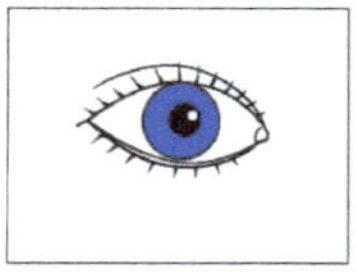

However, the importance of the reversible process to biological science lies beyond these circumstances. This is because all reversible processes, regardless of their nature, be it mechanical, electro dynamical, chemical, or electromagnetic, have one additional constraint that bestows on them their very special place in the biological sciences. It is the Principle of Least Action. And what makes it so indispensible in biological systems is the precision with which it specifies from among all of the possible 'next steps' that the reversible process could evolve to, the one 'next step' that results in the *least action*. And if the process remains in its reversible state after this step, there is again only one 'next step' that the process can evolve to, and it too, is determined by the Principle of Least Action. In this way, we see how any process that remains in a reversible thermodynamic state, must trace out a very specific temporal evolution that we might refer to as the Least Action Process. It does not matter that the process is deterministic, as is the case in DNA transcription, or stochastic, resulting from an ion imbalance at the cell surface. As long as the overall process evolves within the framework of reversible thermodynamics, every 'next step' is precisely determined by the Least Action Principle, and at least in theory, its complete temporal evolution is deterministic.

Many contemporary physicists consider reversible processes to be rare and not very important in the real world. However, when we consider life to be a wholly reversible thermodynamic state, as we will do in this book, we add a nuance of profound importance that brings precision and exactitude to biology that rivals that which we normally reserve to physics.

The above-described treatment is allowed by the Second Law. But, it is not immediately apparent how net reaction velocities approaching this zero limit are achieved, nor is it clear how this unstable, far-from-equilibrium condition can be maintained indefinitely. And yet, this awkward result where the forward and backward progression of time become exactly equal (and in a stable way), and where time appears to stand still, is effectively what happens in living systems. It and its long-term stability, is what gives living systems their ability to self organize.

We will return here in a moment to explore ways of achieving negentropic and reversible process conditions without violating the Second Law. But first, we must make a small detour to understand how events are perceived. We will then return to the more ominous problem of taking an end run around the Second Law of Thermodynamics.

An Atom's Perception of Time

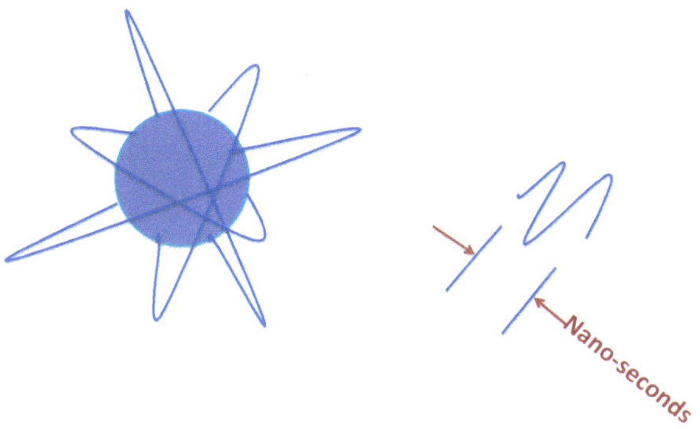

Let's do a thought experiment. Pretend that you are an atom. Your perception of events occurs at the micro-scale of 'atomic events' with the frequency of perhaps, 10^9 per second, or in terms of event intervals, 10^{-9} seconds. Your concept of time is on that scale, and observable events occur in that time frame alone. The reason for this is simple. This is the time scale on which relevant stimuli, and your ability to perceive them, occurs. Time appears to be running very rapidly for an atom. You are seeing the passage of every nano-second.

A Molecule's View Of Time

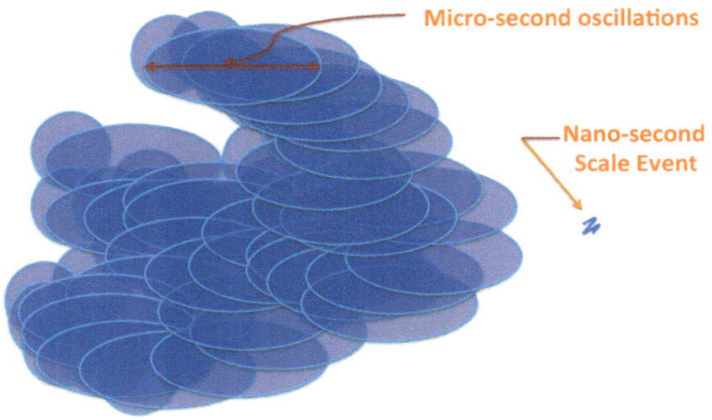

Now look what happens as you, the atom, becomes part of a large molecule. Your perception of those same nano-second events decreases because now the atomic event becomes a smaller part of your experience, and the slower molecular events, such as collision with another large molecule, become more dominant in your field of perception.

Organisms are at the extreme end of this continuum. The atomic event occurring in nano-second time frames doesn't even register in our perception. All that we see are events that are in the time scale of fractional seconds or larger. Our size has made it possible for us to perceive on the second/hour/day/year time scale, something that would have been impossible as that tiny atom. But more important for our purposes here, we have slowed the forward passage of time, and in so doing, made the neg-entropic time direction more dominant relative to the nano-second time scale that the rest of nature operates at.

Organism Perception of Time

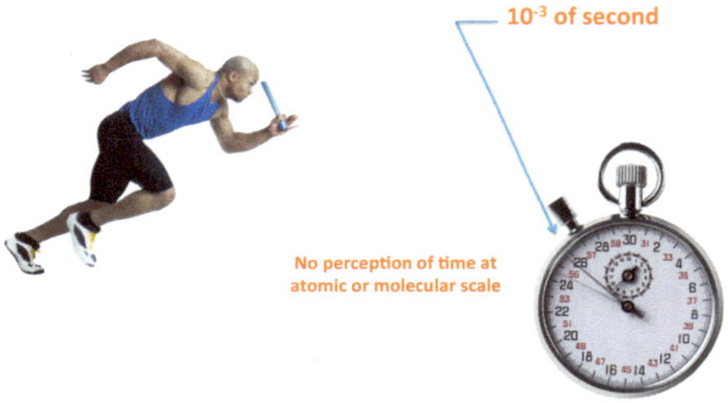

Look at how we have accomplished this change of time frame. We have simply gotten larger. As an atom our kinetic energy was $K.E. = m_{atom}v^2/2$. This is the allowed kinetic energy per particle at the prevailing temperature in accordance with Boltzmann's theory. If we now keep that 'energy per particle' quantity constant, and increase our size 100-fold, our motion must decrease 10-fold:

$$K.E. = 100m_{atom}(\tfrac{1}{10}v)^2/2 \ = 100m_{atom}(\frac{0.1\ m}{sec})^2/2 \ = 100m_{atom}(\frac{m}{10\ sec})^2/2$$

Other ways of looking at this same thing is a decrease in the spatial scale of events by 90% to $\frac{1}{10}$, or an increase in the time scale by a factor of 10. Thus, as long as the temperature is constant, the 100-fold increase in mass translates to perceiving on a time scale where the dominant scale of observables is increased from 10^{-9} seconds to roughly 10^{-8} seconds. To achieve perception in the time scale of 10^{-2} seconds (organism scale), and using this same model, the corresponding observer size approaches $10^{14}m_{atom}$, or about 2×10^{-9}gms as carbon. A typical E-coli weights about 1 pico-gram.

Now let's return to the question that brought us here: How does life make the forward and backward reaction velocities essentially equal without violating the Second Law? We begin by affirming that the velocities and time scales that we have been talking about thus far are the averages, or most probable values resulting from large ensembles of events. The laws governing such processes are statistical in nature, and are described by the laws of statistical thermodynamics.

Now let's do another thought experiment, and this time include a generalized biological enzyme in our calculations. If we now form hydrogen peroxide using the reaction kinetics for our chemical event, the rules are radically different. We will begin by assuming that the active site of our enzyme is occupied by a zinc ion surrounded by at least two His- groups having available

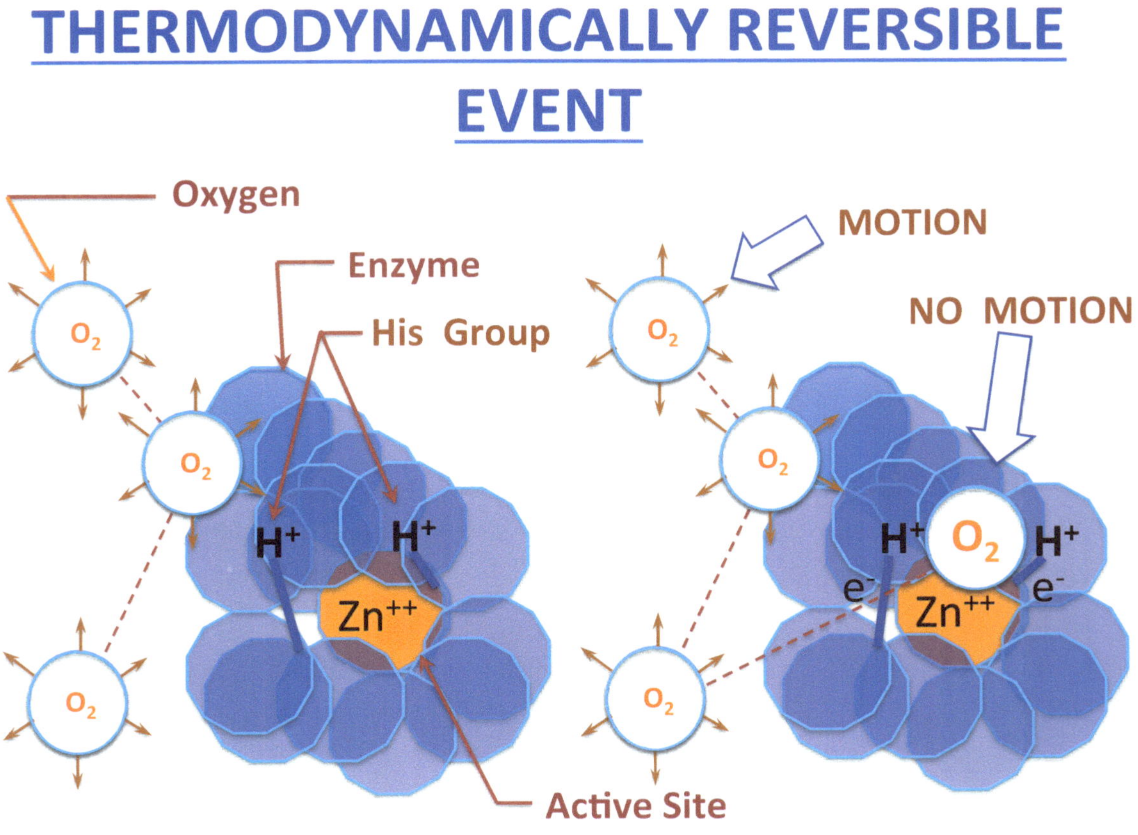

hydrogen atoms in close proximity to the active site. An oxygen molecule approaching this active site has a kinetic energy that is one part of the system's total energy. It is this kinetic energy, which allows it to move into the vicinity of the enzyme where it is attracted either electro-statically or by chemical potential to form zinc oxide. Once this is accomplished, there is no residual molecular motion in the oxygen or its ZnO product. That portion of the system's kinetic energy has been quieted…but not lost. It has not been transferred elsewhere in the kinetic energy pool, nor has it contributed to kinetic energy in the enzyme molecule. It has been absorbed into the oxygen-enzyme complex.

Under these circumstances, immobilized H_2O_2 can form *without any loss of free energy to 'energy of motion'*, and we understand this reaction to be in its thermodynamically reversible state. It has identical probability of going in the entropic and neg-entropic time directions, and thus, its forward and backward reaction rates are identical. In essence, we have taken the hydrogen peroxide reaction out of the domain of statistical thermodynamics where the kinetic energy of constituent parts contributes to uncertainty, and placed it entirely within the very precise realm of determinism, where the Principle of Least Action alone, determines the overall process' next step.

This state, where the forward and reverse directions of time passage are essentially equal, places us at the very limit of the Second Law, where processes are thermodynamically reversible, and entropy change is identically zero. Relative to the rest of nature, this process has taken a step backward in time. This is how life achieves neg-entropic character in a world of statistical processes.

C. The Objectives of Theoretical Biology

The theoretical biologist is searching for the living state's first principles, with the secondary objective: applying these to the understanding of life, its permutations, and its diseases. It is a quantitative field of study that is based in mathematical models that are similar to those that one finds in physics and chemistry. The models that we seek should have a precision that is comparable to that found in Maxwell's equations for electricity and magnetism, Gibbs' calculations in dilute chemical solutions, or Helmholtz calculations for chemical and electro-dynamic systems.

The difference between these models and our model of the living state will be in our choice of a theoretical framework; one that is better suited to matter's living state. In particular, we will be using reversible thermodynamic calculations to deal with the fundamentals of the living state. We will employ irreversible thermodynamics only at the living system's boundary conditions, or where disease or other aberrations to the cell's energy structure interfere with the reversible thermodynamic foundations of our model.

One of the things that we will be discovering about living systems is that they are very exact in their forward progression. However, their response to experimental conditions will tend to be highly variable. This is not because of randomness in the living system's physics or chemistry. This cannot occur because Laws of Nature have no randomness. Therefore, the application of physical and chemical principles in our model must remain very precise. In fact, as we explore this subject in its most intimate details, we will come to understand that life's variability is highly systemized, and has its roots in the organism's exact response to a highly variable environment.

Objectives of Theoretical Biology

1. **We need to understand life as a stable, far-from-equilibrium state of matter.**
2. **What is the mechanism of energy accumulation and storage in this thermodynamically reversible systems?**
3. **What is free energy? How is it ordered in matter's living state**
4. **How can vast amounts of accumulated heat energy exist in a room temperature system?**
5. **We need to determine the living systems fundamental forcing function(s).**
6. **How is the precise sequencing of cell synthesis and metabolic events controlled?**

Let us begin by laying out some objectives for our study of theoretical biology. Because this is a new discipline, our focus will be limited to only six fundamentals:

1. Biological systems are spatially localized regions of neg-entropy. We see this in the very organized molecular structures within the cell, and on a larger scale in the useful work that living cells and organisms can accomplish. We want to understand how stable far-from-equilibrium conditions can be maintained, and how they function at the level of physics.
2. We want to understand the source of the energy that creates this far-from-equilibrium state. Nutrition and digestive processes behave according to irreversible thermodynamics, putting them outside the range of possibilities for this energy's immediate source. We will seek to

understand energy accumulation at its first encounter with our thermodynamically reversible, living system.

3. We want to understand how the energy in a living system is organized and stored. To get there we will have to ask: "What is free energy within the context of a living cell?" This storage must be every bit as far-from-equilibrium as the molecular quantities. We will be looking for methods that nature might employ to store energy in a very stable, far-from-equilibrium way.
4. It will also be important for us to reconcile how extremely large quantities of 'heat energy' can be stored in a cellular system that appears to be at room temperature. This energy storage manifests when we burn the cell.
5. However, the most elusive element in our quest for a mathematical model of a living cell will be the forcing function that propels cell structure and function in the positive time direction. More succinctly, we want to know what makes the cell 'go'.
6. Finally, we want to know how biochemical events are sequenced to produce exact outcomes, even in a sometimes hostile (random) environment.

The objective of this manuscript is to present a theoretical framework for understanding the structure and function of living systems. We will move toward this objective with an emphasis on a single animal cell. However, the principles involved in this discussion will be much more universal, having applications to both plant and animal cells, and ultimately applications to complex organisms like ourselves. Our journey will take us into areas of physics where we will study energy storage and exchange, at its most fundamental levels. This will, in turn, transport us into the realm of thermodynamically reversible processes and to the very limits of the great entropy principle.

You might want to look at this as a vacation to an entirely different world, one where the very laws of nature perform in new and unexpected ways. We will see how energy accumulation and storage in our cell, operates according to completely different rules than those that we are familiar with in matter's non-living state. The very concept of time will take on new meanings.

D. The Modeling Dilemma

Finally, we will need to explore some of the problems in developing a deterministic model for a system that is known to be highly variable, almost with a stochastic overlay. Our studies in biology show us how the cell is, in general, a highly deterministic system. There is extraordinary fidelity in its production of daughter cells, and progeny. And yet, when we measure biological response in our laboratories, we find a degree of variability that is strikingly at odds with the precision that we expect. How is it, that this very precise machine exhibits the kind of variability that sometime resembles the toss of a dice? And yet, even in the face of this variability, the system maintains its extraordinary precision.

Challenges in Biological Modeling

- **We need to separate deterministic and stochastic processes. How can each be modeled?**
- **Can we work with the large number of molecular variables? How can we simplify?**
- **What controls the time sequencing and precision of biochemical events?**
- **What constitutes a calibration data set?**
- **What makes the living cell 'go' forward in time?**

A second modeling issue is of equal import. We are facing an extremely complex system that contains an extraordinarily large number of different molecules. Persons experienced in mathematical modeling will quickly point out the need to simplify the dimensionality of the system by several orders of magnitude, to even begin the modeling effort. But what do we eliminate? We could compromise our ability to model the whole by throwing away its important element(s), or by not giving it/them the attention that they deserve. More to the point, it would be senseless to attempt initial modeling efforts at the level of detail that is required to model the molecular quantities. This is simply not possible. We must start simple, and then advance toward this more ambitious goal. Eventually, we will get there.

The modeling approach taken here is unique. Instead of modeling molecular quantities, I have chosen to eliminate them altogether, and instead focus all of my effort on studying the energy structure within the cell. When I use the words 'energy structure' in this context, I am not talking about the individual energy requirements for specific covalent bonds, or of specific biochemical reactions. Instead, we will be looking at the way that the cell's total energy is organized in relationship to one or two fundamental variables. For example, if we are storing energy as photons, we might choose to order them by their frequency, and then according to where they might occur in the cell's structure. We might even choose to organize the energy quanta according to their temporal placement in the mitotic cycle.

The idea here is to build a scaffold for more detailed modeling efforts at the molecular level. We might, for instance, locate specific stages in the cell cycle (S1, G, M), or specific metabolic functions (respiration, glycolysis) on such a scaffold. This approach has the advantage of being very simplified, having only three or four variables. It is hoped that by knowing the structure of

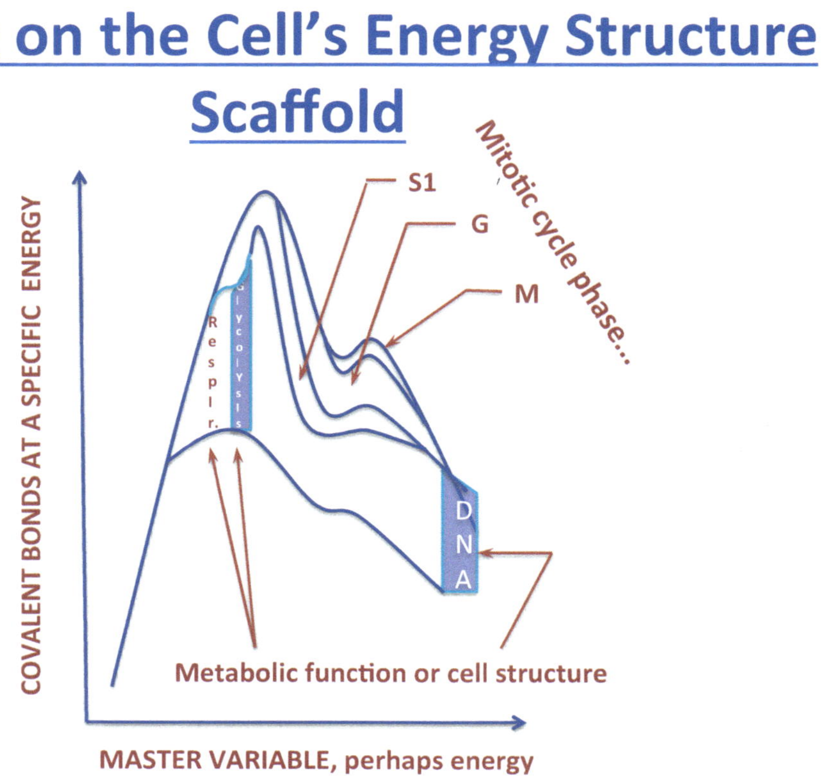

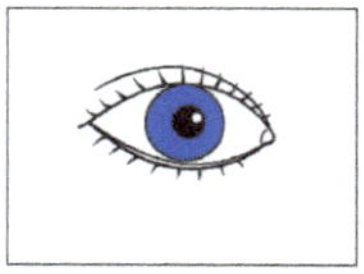

the systems energy, it will then be possible to isolate within that energy structure, specific cellular structures and cell functions, and then build more detailed molecular models onto this energy scaffold. In the next chapter, we will lay out a theoretical framework for these far-from-equilibrium energy calculations, and arrive at a candidate form for the required energy scaffold.

Chapter 2

A Theory of Far-From-Equilibrium Heat Storage

A. Introduction

Planck's blackbody emittance equation (1) is the universally accepted model for heat radiation's equilibrium, spectral distribution. It has been found superior to any other contemporary form (2,3,4,5,6). However, this acceptance is only justified for equilibrium, and leaves two important issues unresolved. First, Planck's solution provides no insight into non-equilibrium or far-from-equilibrium states, or the mechanisms of redistribution between equilibrium states (7). Only Ehrenfest (8) has explored redistribution mechanisms, and Forte (9) describes a non-equilibrium Wien Displacement Law.

Secondly, Planck's energy quanta violated the continuity requirements of Maxwell's equations. Einstein first enunciated this discrepancy, and Planck spent the next 2 decades, unsuccessfully trying to resolve it. Meanwhile, the experiments of Stark (10), and Einstein's light particle theory (11) demonstrated the dual nature of light, galvanizing the discontinuity's place in physics. More recent studies of non-quantum blackbody theory (12,13,14,15,16,17) have not reconciled this conflict.

This paper explores one possible avenue to a non-equilibrium blackbody equation. The goal here, is understanding free energy partitioning between the domains of heat radiation, and molecular motion. One of the model's solutions is then interpreted as an avenue to understanding far-from-equilibrium energy storage in living cells and cold fusion devices.

B. Theory

Planck (18) viewed the separation of all physical phenomena into reversible and irreversible processes as the most elemental, and most important, because all irreversible processes share a common similarity that makes them unlike any reversible process. This distinguishing characteristic is the transformation of heat energy to motion, which can in no way be referred back to the process from which it came. This research considers Maxwell's electromagnetic wave traveling undiminished in time, its information content preserved, until it encounters a material particle.

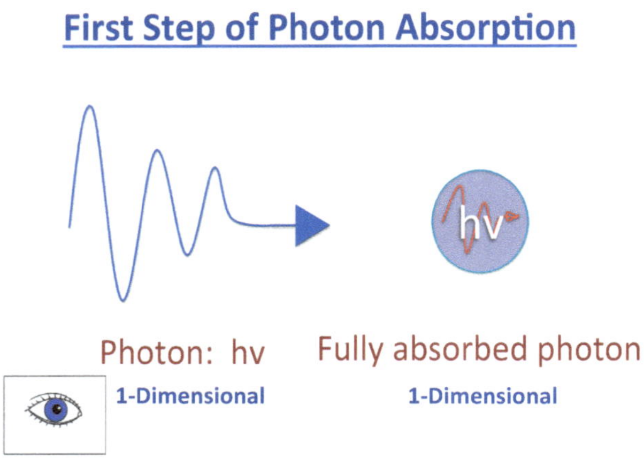

Light absorption is considered a two-step process. The first is an adiabatic reversible step, wherein one-dimensional light energy is absorbed in a quantum amount, $h\upsilon$, by an electron, and is wholly contained within it. The absorbed quantum is still 1-dimensional (1D), remains within the domain of reversible thermodynamics, and does not emit Joule heat. There is no recourse to the Second Law during this first step.

The absorption process' second step is a dimensional restructuring that the 1-D electrical quantum undergoes in evolving into its 3-D equivalent, electrical charge density. This occurs in accordance with the Equip-partition Theorem, along the axes of the electron's three spatial coordinates. The resulting displacement of the generalized coordinates translates to 3-D motion, the evolution of Joule heat, and irreversibility. The magnetic vector has no 3-D equivalent, and can only transform to 1-D paramagnetic spin. Accordingly, photon de-coupling distorts time's fabric, giving rise to the characteristic spectral emittance.

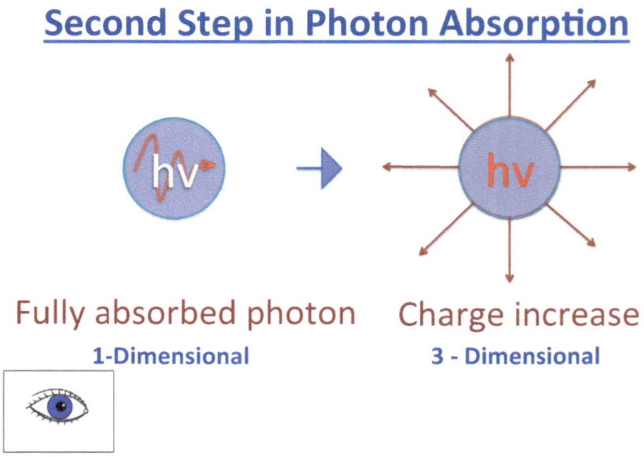

LIGHT ABSORPTION BY AN ELECTRON

(TWO STEP PROCESS)

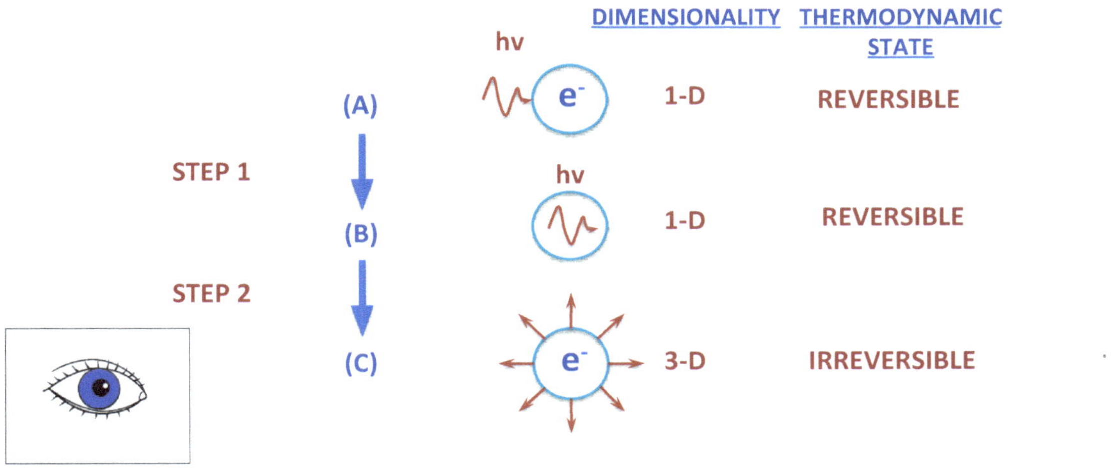

This second absorption step is represented by integrating the Dirac delta over the 3-D transformation's time interval

(1) $$\int_{t1}^{t2} \frac{1}{t}\,dt = \ln\left(\frac{t_2}{t_1}\right) = \ln\left(\frac{\upsilon_1}{\upsilon_2}\right) \left[\textit{frequency domain}\right]$$

with variance distribution:

(2) $$\sigma^2_{\upsilon_1/\upsilon_2} = \left[\ln\left(\frac{\upsilon_1}{\upsilon_2}\right)\right]^2 = f_1\left(\frac{\upsilon_1}{\upsilon_2}\right)$$ 1st Integral

The probability distribution's re-structuring, ν_m from a 1-D quanta, $h\nu$, to three dimensions requires taking the next two integrals of the variance function relative to the Wien frequency, ν_m. Thus:

(3) $$f_n\left(\frac{\upsilon_1}{\upsilon_m}\right) = \iint \left[\ln\left(\frac{\upsilon_1}{\upsilon_m}\right)\right]^2 d\left(\frac{\upsilon_1}{\upsilon_m}\right)$$

The results are summarized as:

(4) $$f_2\left({}^{v_1}\!/_{v_m}\right)=\left({}^{v_1}\!/_{v_m}\right)^1\left[\ \left(\ln\left(\frac{v_1}{v_m}\right)\right)^2-2\ \ln\left(\frac{v_1}{v_m}\right)\ +2\right]$$ 2nd Integral

(5) $$f_3\left({}^{v_1}\!/_{v_m}\right)=\left({}^{v_1}\!/_{v_m}\right)^2\left[\frac{1}{2}\left(\ln\left(\frac{v_1}{v_m}\right)\right)^2-\frac{3}{2}\ \ln\left(\frac{v_1}{v_m}\right)+\frac{7}{4}\right]$$ 3rd Integral

The third integral represents all of the possible interconnections between any arbitrary frequency and the Wien frequency. $\nu_m$is the most probable frequency at the prevailing temperature. ν_1 is the damped frequency continuum of the blackbody spectra.

Assuming the radiation absorption function to be exponentially distributed,

(6) $$a_v = e^{f_3(v_1/v_2)}$$

and substituting this absorption and the Raleigh-Jeans emittance into Kirchhoff's Law:

(7) $$K(v_1) \quad = \quad \frac{e_v}{a_v} \quad = \quad k_b T_m \cdot \frac{v_1^2}{c^2} \quad \cdot \quad \frac{1}{e^{f_3(v_1/v_m)}}$$

Blackbody =	Emittance /	Absorptance
Spectra	(Rayleigh Law)	(This study)

This spectral distribution: 1) is derived entirely from classical theory; 2) contains the discontinuity indirectly, $\nu_m(h)$; 3) incorporates the consequences of electro-magnetic theory (Rayleigh-Jeans Law); and 4) suggests a mechanism for exchange of energy between frequencies. Sears [19] gives Wien's Displacement Law as:

(8) $$\nu_m = 5.89\times10^{10} \cdot T_R(^oK) \quad \text{(Wien frequency)}$$

Figure 2 displays features of the blackbody radiation spectra described in this way. The figure also displays calculations using Planck's Equation:

(9) $$K'(\nu_1) = h\nu_1 \frac{\nu_1^2}{c^2} \cdot \frac{1}{e^{\frac{h\nu_1}{k_b T}}-1}$$

The agreement is close, but not exact, differing by less than 2% at ν_m, and in the 5%-8% range to the left, where Eq. (7) better represents Raleigh-Jeans. The curves terminate at the point where the calculations yield partial quanta. The number of quanta is obtained by dividing the spectral emittance by the conversion factor υ^2/c^2 and then dividing by $h\upsilon$.

Steady State Blackbody Spectra

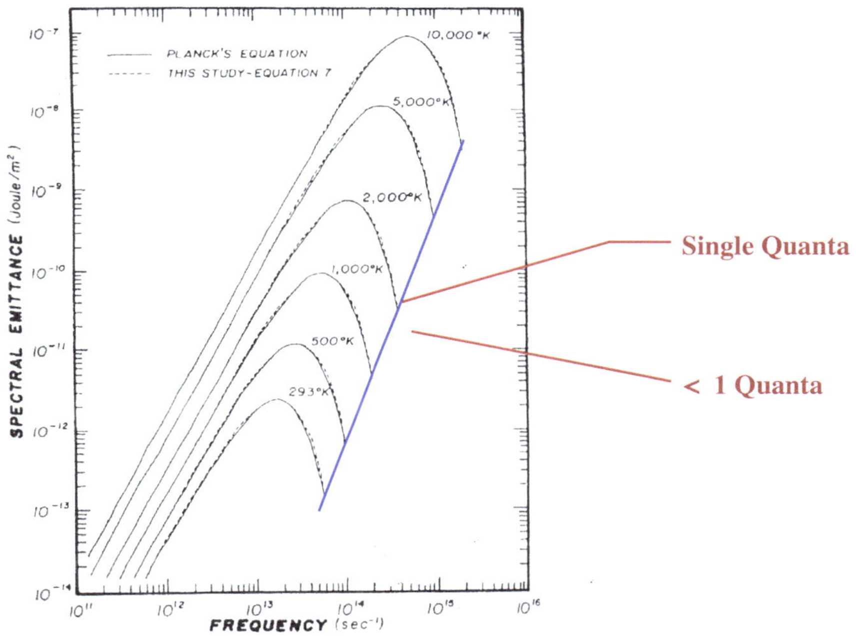

C. Discussion

Both the two-slit experiment and the photoelectric effect are consistent with this theory. The wave properties of light are unaltered. The theory eliminates discontinuity as a property of light, placing it instead in the electron's absorption of light, or more precisely, in the electrons absorption of light only in quantum amounts. This reassignment isn't contrary to, nor does it change, existing theory.

Eq. (7) offers two significant advances over Planck's, which are instructive in furthering our understanding of heat processes. The first is Eq. (7)'s explicit statement for energy transference between frequencies. This was identified at the outset as the distinguishing characteristic of the required non-equilibrium blackbody form. Eq. (5) suggests that the common channel for energy re-distribution is the Wien frequency, since each spectral frequency is explicitly related to it. Planck's equation can also be shown to contain the same ratio (21).

Second, Eq. (7) contains two distinct thermodynamic scales, representing the entire range of non-equilibrium heat conditions. The concept of two temperature scales is not new (22,23,24,25,26,27). The first of these scales is the classical thermodynamic temperature, of the Rayleigh-Jeans Law, T_m . It is common to both equations, and expresses the temperature of thermal motion alone.

Non-Equilibrium Blackbody Equation

$$K(\upsilon_1) = \frac{e_\upsilon}{a_\upsilon} = k_b\, T_m \cdot \frac{\upsilon_1^2}{c^2} \cdot \frac{1}{e^{f_3(\upsilon_1/\upsilon_m)}}$$

Where: $$f_3\left(\upsilon_1/\upsilon_m\right) = \left(\upsilon_1/\upsilon_m\right)^2\left[\frac{1}{2}\left(\ln\left(\frac{\upsilon_1}{\upsilon_m}\right)\right)^2 - \frac{3}{2}\ln\left(\frac{\upsilon_1}{\upsilon_m}\right) + \frac{7}{4}\right]$$

$\upsilon_m = 5.89 \times 10^{10}\, T_R\ (^oK)...WeinFrequency$

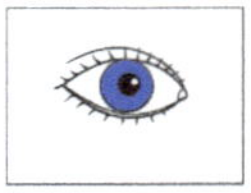

T_m = Thermodynamic Temperature

T_R = Radiation Temperature, ***A New Temperature Scale***

The second temperature, that contained in the Wien Displacement Law, is identical to the first where the system is in equilibrium. However, it is fundamentally different from T_m in ways that could give profound meaning to Eq. (7). This is the radiation temperature, T_R. That it can be expressed in the same units as the classical thermodynamic temperature is seen in the equilibrium case. However, changes in T_R, independent of the thermodynamic temperature, shift the spectral distribution in plausible non-equilibrium ways that may provide insight into both non-equilibrium and far-from-equilibrium heat processes.

In the next figure, T_R, and consequently the Wien frequency, remains constant while T_m increases from $300\,^\circ K$ to $10^5\ ^oK$ (Case A). This represents a sudden frictional input of heat to a material body that is initially at thermal equilibrium. Similarly, T_R can be increased without a corresponding increase in the thermodynamic temperature (Case B). The radiation density within the blackbody is increased without a corresponding increase in the Rayleigh-Jeans emittance. The new region delineated by this spectral distribution consists primarily of higher energy radiation, but the process from which it arises appears to an observer to be adiabatic, and might therefore, be viewed as completely reversible. From this theory's standpoint, the energy content within this new region (Case B) consists entirely of radiation transfers that are undergoing the

FAR-FROM-EQUILIBRIUM BLACKBODY SPECTRA

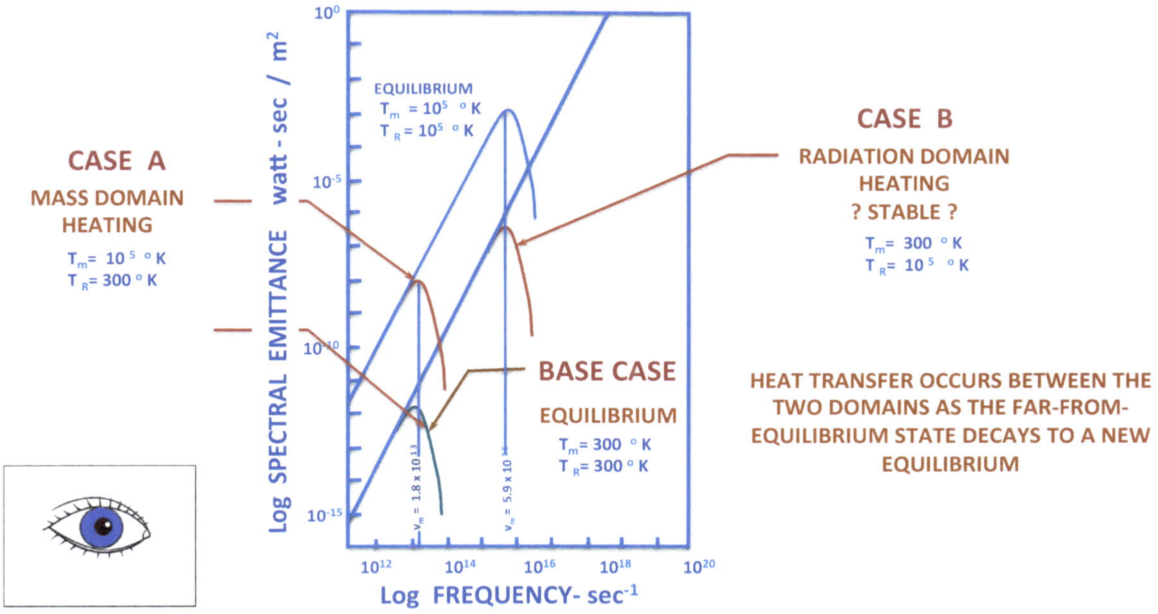

first stage of radiation absorption, alone. That is, radiation is fully absorbed in its one-dimensional form and immediately re-emitted. There is no de-coupling of light's electro-magnetic structure, and therefore no entropy increase. This is the initial condition when high-energy radiation strikes a body initially at equilibrium.

Taking this result further, one might ask: Are there states in nature that exploit the energy/entropy relationship suggested by these calculations? There might be. Living systems are constructed of high energy covalent bonds that both, represent very far-from-equilibrium conditions, and store larger amounts of electro-magnetic energy than would normally exist at the T_m. It is possible that Case B shows how far-from-

Reversible Photon Resonance Between Electrons Within Covalent Bonds

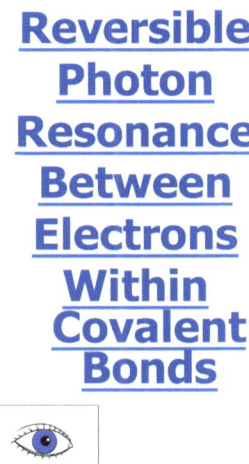

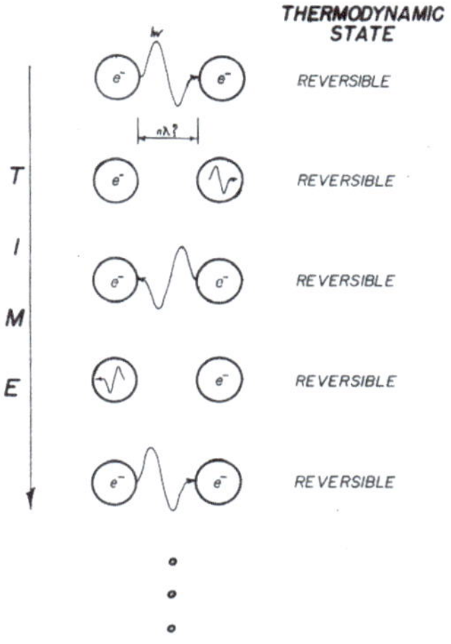

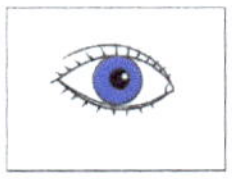

equilibrium energy storage might be masked from ambient thermodynamic conditions in matters living state. Each covalent electron pair shares the wave function, ψ^2, alternately absorbing and re-emitting light energy, but only in a manner consistent with this theory's first absorption step. This portion of the heat radiation spectra is localized (masked) between electron pairs, and does not contribute to either the measurable heat spectra or to dielectric losses. Thus, the thermodynamic temperature of the cell (T_m) is unaffected, and a stable far-from-equilibrium condition with lower localized entropy, is possible. The degree of entropy decrease is defined by the separation between T_m and T_R. The permanence of that change appears to depend on irreversible storage of neg-entropy outside mechanistic pathways back to equilibrium (28). As the energy storage requirements begin to exceed the capacity of the covalent bond system, other mechanisms for storing this biological energy in a stable far-from-equilibrium state might be found in covalent bond, and probably even excited nuclear states, as Mossbauer resonance. Eq. (5) suggests enormous capacity for far-from-equilibrium entropy absorption and the information storage this implies.

A second example of where this type of heat theory may prove important to scientific understanding is found in what has been called cold fusion. Energy storage in a palladium or nickel electrode might occur in the radiation domain (Case B in Figure 3), first as excited electronic states and a corresponding increase in the redox state, and later in excited nuclear (Mossbauer) states. Is it possible that such a system could exhibit radiation temperatures approaching $10^7\ {}^0K$ while the ambient temperature of the electrolysis apparatus hovers around $330\ {}^0K$?

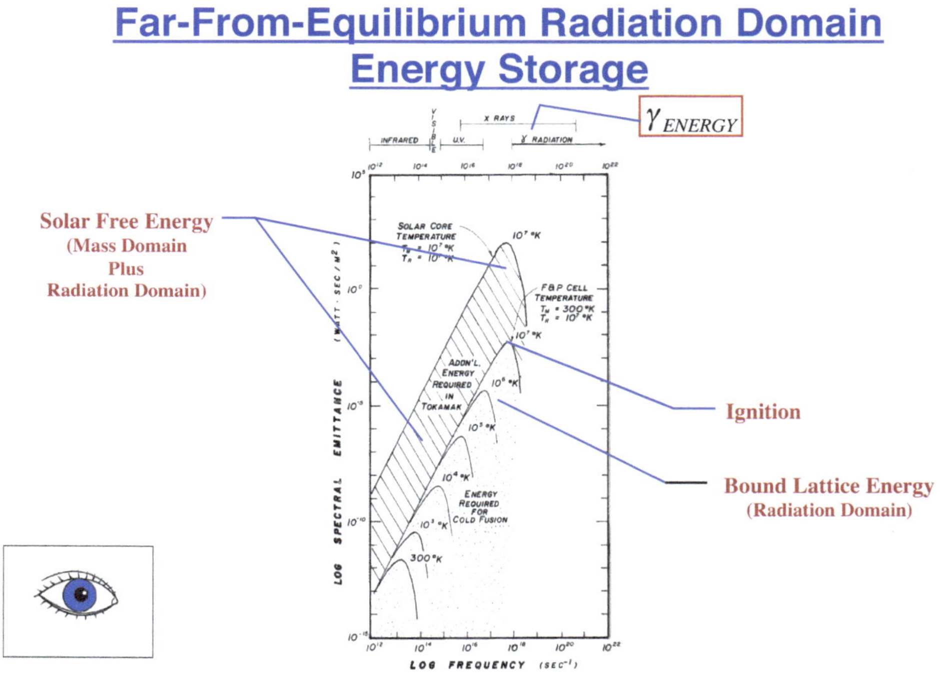

Stranger paradigms have occurred in the history of science.

A third example of far-from-equilibrium spectral energy might be found in sono-luminescence, wherein mechanical energy is converted to electro-magnetic energy. In this case, mechanical energy increases T_m instantaneously without a corresponding increase in T_R (Case A in Figure 3). Lacking any mechanism to maintain this far-from-equilibrium condition, the system spontaneously moves toward equilibrium by channeling the stored mechanical energy through the Wien frequency channel, and thence, into the radiation domain. If the energy flux is high enough, visible light is observed.

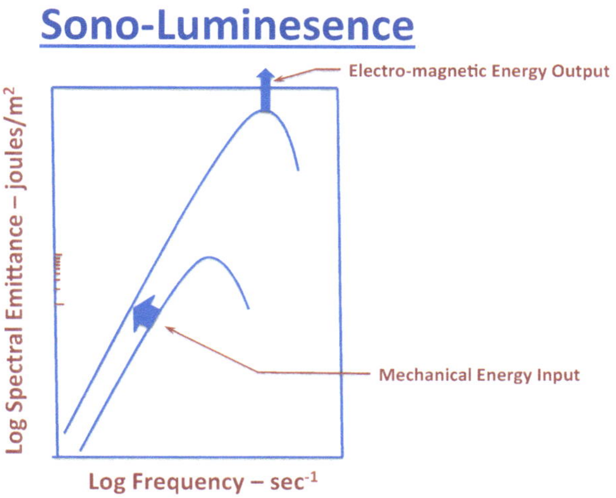

The 1-D to 3-D transform function given by Equation 5 could possibly be a mathematical statement of the Second Law at the boundary between electrodynamics and mechanics. In its temporal form (19) the equation represents the relative dominance of the forward (entropic) and backward (negentropic) reaction directions. The entropic direction is that in which waste heat of motion is produced, while the negentropic direction is characterized by increased order where the radiation domain predominates.

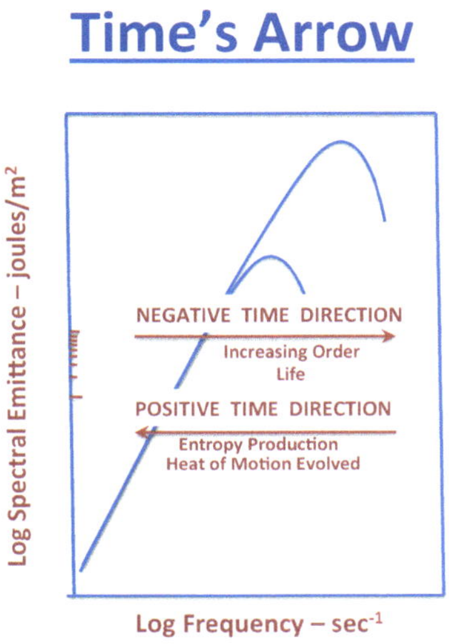

References

(1) M. Planck, Verhandlunger der Deutschen Physikalischen Gesellschaft, 2, 237, (1900), or in English translation: Planck's Original Papers in Quantum Physics, Volume 1 of Classic Papers in Physics, H. Kangro ed., Wiley, New York (1972).

(2) H. Rubens and F. Kurlbaum, Ann. Physik, 4, 649 (1901).

(3) F. Paschen, Ann. Physik,4, 277 (1901).

(4) E. Warburg, Ann. Physik, 48, 410 (1915).

(5) W. Nernst and T. Wulf, Ber. deut. phys. Ges., 21, 294(1919).

(6) W.W. Coblenz, Dict. Appl. Phys., Vol. IV, "Radiation".

(7) T.S. Kuhn, Blackbody Theory and the Quantum Discontinuity 1894-1912, Oxford University Press, New York (1978).

(8) Kuhn found references to non-equilibrium transitions only in Ehrenfest's notes.

(9) J. Fort, J.A. Gonzalez, J. E. Liebot, Physical Letters A, 236,193-200 (1997).

(10) J. Stark, Phys.ZS.,8(1907), 913-919, received 2 December 1907.

(11) A. Einstein, Ann. d. Phys.,17(190), 132-148, 1905.

(12) H. Dingle, Phil Mag, XXXVII,246, 47 (1946).

(13) T.H. Boyer, Phys. Rev. 182, 1374 (1969).

(14) T.H. Boyer, Phys. Rev. 186, 1304 (1969).

(15) T.H. Boyer, Phys. Rev. D, 11, 790 (1975).

(16) T.H. Boyer, Phys. Rev. D, 29, 1089 (1984).

(17) D.C. Cole, Phys. Rev. A, 45, 8471 (1992).

(18) M. Planck, Eight Lectures in Theoretical Physics-1909, translated by A.P. Wills, Columbia U Press, NY (1915).

(19) F.W. Sears and M.W. Zemansky, University Physics, 3rd ed., Addison-Wiley, Reading, MA (1964).

(20) See T. Preston, Theory of Heat, Macmillan and Co, London (1929) for a description of their methods and results.

(21) If we rearrange Equation (8) and substitute for T in Planck's Equation (9), the exponential term becomes $e^{2.8276\left(\upsilon_1/\upsilon_m\right)}$.

(22) B.C. Eu, L.S. Garcia-Colin, Phys. Rev. E, 54, 2501 (1996).

(23) D, Jou, J. Casas-Vazquez, Phys. Rev. A, 45, 8371 (1992).

(24) K. Henjes, Phys. Rev. A, 48, 3194 (1993).

(25) W.G. Hoover, B.L. Holian, and H.A. Posch, Phys. Rev. A, 48, 3191 (1993).

(26) D. Jou, J. Casas-Vazquez, Phys. Rev. A, 48, 3201 (1993).

(27) J. Fort, D. Jou, and J.E. Leobot, Physica A, 269, 439(1999).

(28) G. Nicolis, I. Prigogine, Self-Organization in Non-Equilibrium Systems, John Wiley and Sons, NY, 1977.

Chapter 3

Cell Structure and Function - The Synthesis Pathway

A. Introduction

The physical principles underlying entropy relationships in matter's living state continue to baffle scientific inquiry. Despite our advances in understanding the complexity of cellular biochemistry and genetics, the physical laws that give rise to life seem to be beyond what the known laws of physics allow. This of course is not true; and in spite of arguments favoring principles outside our understanding to account for matters living state [1], there is nothing that contradicts the view that life results from extensions of existing physical laws that have not yet been coalesced into the proper theoretical framework.

"The methods of physics and chemistry are ideally suited for dealing with homogeneous classes with their interchangeable components. But experience shows that the objects of biology are radically inhomogeneous both as systems (structurally) and as classes (generically). Therefore, the method of biology and, consequently, its results will differ widely from the method and results of physical science."

Walter M. Elsasser, *Atom and Organism: A New Approach to Theoretical Biology* (1966)

Schrodinger was instrumental in explaining how the principles of physics might be applied to living systems. His insightful work [2] broke down the complexity of what was known about the cell in 1941, into concepts that could be isolated and generalized. He was among the first to describe the living state's negentropic character, its apparent ability to locally decrease entropy.

At about that same time, Szent-Gyorgyi [3] was coming to the understanding that ensembles of proteins might possess common energy bands similar to those being studied in semiconductors. He based his supposition on studies showing that large numbers of similar enzyme molecules appeared to be acting in concert to effect biochemical transformations. His belief was supported by studies on chlorophyll, urease, and fumarase among others. But it went farther, even suggesting that this type of behavior might go beyond electron tunneling among similar molecules, to perhaps, discrete energy levels within the entire cell structure. He speculated that electrons might be directed over large distances, allowing them to fall to lower energies only where they would perform useful work.

With characteristic eloquence, Monod [4] summarized the state of enzyme form and function in 1971. He wrote of the remarkable structural fidelity, precise globular forms, and unerring specificity of the enzymes that were the subject of his life's work. In the end, he concluded that something beyond genetically determined peptide sequence imparted the information content found in their condensed form. What exactly, he could not say.

That unfortunate circumstance is unchanged even today. The only thing that we can say with any certainty is that somehow within the context of the cellular environment, the enzyme's amino acid sequence acquires large amounts of information (negentropy), which ultimately specifies both its form and a very specific function.

The biochemical data that has been amassed during the last 60 years characterizes with extraordinary precision the mass quantities within the cell. DNA, RNA, proteins, and countless other molecules have been described in both their structure and function. Their mutual interactions and their interactions with other molecular forms will soon be exhaustively catalogued. Yet it doesn't seem that we are that much closer to understanding the principle(s) that cause a linear amino acid sequence to acquire its extraordinary negentropic character, much less, nature's closely guarded secrets of the living state.

THE BIOLOGICAL MODELING DECISION

MOLECULAR QUANTITIES	COVALENT BOND ENERGIES
• HIGH INFORMATION CONTENT	• NO APPARENT INFORMATION CONTENT
• WE CAN SEE AND DESCRIBE THEM	• REALITY WITHOUT MEASUREMENT
• TOO NUMEROUS TO MODEL	• TWO OR THREE VARIABLES

It is this author's belief that this state of affairs reflects our continuing emphasis of life's molecular forms. This is only natural. We can see them. But, still they are only about half of the picture. The unseen half, consisting of energy fields that accompany the molecular forms and hold their covalent bond structure in a far-from-equilibrium state, are surely as important. I am not referring here to the individual energy relations in any specific molecule, but rather the overall context of the cell's covalently bonded, electro-magnetic field structure.

We should expect this electro-magnetic signature to be very precise. After all, the 'glue' that holds the atoms and molecules together should reflect the far-from-equilibrium state in those quantities. But, more to the point, it should have its own negentropic character, which once revealed, may tell us even more about life. Since the answer that eluded Schrodinger, Elsasser, Monod, Luria [5], and countless other extraordinary intellects has not yet been evident in the mass quantities, perhaps it is the radiation field that possesses Monod's *ultima ratio* [4].

This paper is one small step in that direction. It begins with some thoughts on non-equilibrium heat theory, and speculates on the spectral distribution of heat energy within living systems, and the nature of its far-from-equilibrium condition.

B. The Dual Nature of Heat Energy

The amount of heat energy stored in any physical system is the sum of two quantities: the energy stored in its heat radiation field, and that contained in thermal motion. The first quantity can be thought of as existing entirely in the electro-magnetic domain (radiation domain), while the second exists only in the domain of material particles (mass domain). Energy exchange between these two domains occurs continually, always driving the system toward a maximum entropy condition that an observer sees as thermal equilibrium. It was Helmholtz's great contribution to physics to recognize that heat energy ultimately reduces to motion. And as long as the process is reversible, and the driving force is in the other direction, the reverse can also be true.

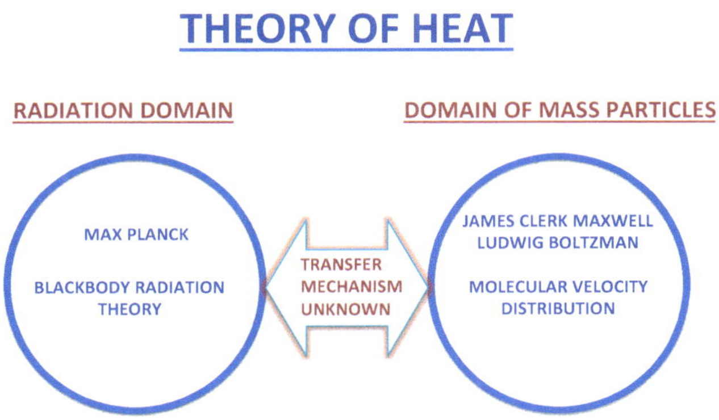

The distinction between these two types of heat energy is well known. We find that when two material bodies are rubbed together, the resulting molecular motion causes both to become heated. It is also possible to impart an identical amount of heat to these same two objects by exposing them to a source of electro-magnetic energy (eg. the sun). And in fact, when the heat energy distribution in both cases has reached an equilibrium, it is not possible to distinguish by which mode the equilibrium was affected. The equilibrium heat radiation spectrum was first described by Planck [6]. Our understanding of thermal motion in a gas at equilibrium is associated with the revered names of Boltzmann [7] and Maxwell [8].

A heat system rarely achieves true equilibrium. It generally perturbates around a quasi-equilibrium state, where microscopic changes predominant, or it evolves inexorably toward a new equilibrium state. A third non-equilibrium condition might also be possible, one in which the heat distribution of the radiation and/or the mass domain is held in a stable far-from-equilibrium state.

C. Non-Equilibrium and Far-From-Equilibrium Heat Energy Storage

I recently proposed an alternative form for Planck's blackbody radiation equation that suggests an avenue to non-equilibrium, and possibly even far-from-equilibrium solutions to the heat radiation problem [9]. The candidate, non-equilibrium blackbody spectral distribution was found to have the form:

(1) $$K(\nu) = k_b T_m \cdot \frac{\nu_1^2}{c^2} \cdot \frac{1}{e^{f_3(\nu_1/\nu_m)}} =$$

Where:

(2) $$\nu_m = 5.89 \times 10^{10}\, T_R \ \ (^oK)$$ (the Wein frequency)

(3) $$f_3(\nu_1/\nu_m) = (\nu_1/\nu_m)^2 \left[\frac{1}{2}\left(ln\left(\frac{\nu_1}{\nu_m}\right)\right)^2 - \frac{3}{2}\, ln\left(\frac{\nu_1}{\nu_m}\right) + \frac{7}{4}\right]$$ (entropy operator)

and: $K(\nu)$ exists where its magnitude equals or exceeds $h\nu$, a single quanta,
T_m is the thermodynamic temperature,
T_R is a new quantity described as the radiation temperature.

This last equation, $f_3(^{\nu_1}/_{\nu_m})$, expresses the interconnectedness between the range of blackbody frequencies, ν_1, and the Wein frequency, ν_m. Several features of the theory behind this equation are relevant to what follows.

The first is the hypothesis that radiation absorption might take place as a two-step process wherein light's one-dimensional form is initially absorbed by an electron in discrete quantum amounts, and then, in the second step, it is partitioned into three-dimensional electrical charge and one-dimensional magnetic spin. The first step is shown to be an adiabatic reversible process; the second: entropic.

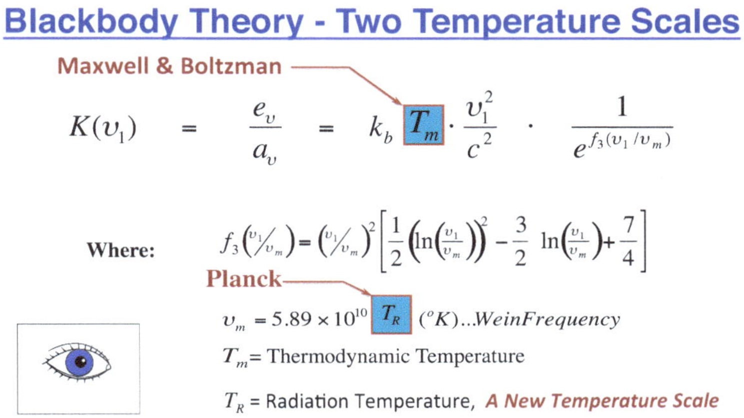

Blackbody Theory - Two Temperature Scales
Maxwell & Boltzman
$K(\upsilon_1) = \frac{e_\upsilon}{a_\upsilon} = k_b\, T_m \cdot \frac{\upsilon_1^2}{c^2} \cdot \frac{1}{e^{f_3(\upsilon_1/\upsilon_m)}}$
Where: $f_3(^{\upsilon_1}/_{\upsilon_m}) = (^{\upsilon_1}/_{\upsilon_m})^2\left[\frac{1}{2}\left(\ln\left(\frac{\upsilon_1}{\upsilon_m}\right)\right)^2 - \frac{3}{2}\ln\left(\frac{\upsilon_1}{\upsilon_m}\right) + \frac{7}{4}\right]$
Planck
$\upsilon_m = 5.89 \times 10^{10}\, T_R\ (^oK)$...WeinFrequency
T_m= Thermodynamic Temperature
T_R = Radiation Temperature, A New Temperature Scale

Secondly, the theory suggests that the radiation and mass domains might be represented by independent temperature scales. The figure illustrates the principle characteristics of the non-equilibrium blackbody radiation spectra. The Base Case illustrates an equilibrium solution for 293 ^oK. It is close to, but not exactly Planck's solution. Case A is a condition where the mass domain temperature alone is increased (friction input of heat), while Case B represents a similar increase in the radiation domain temperature resulting from radiation absorption. These two cases are far-from-equilibrium states that would normally decay to an equilibrium spectrum similar to, but with higher total energy than, the Base Case. At equilibrium the radiation and mass temperatures become identical.

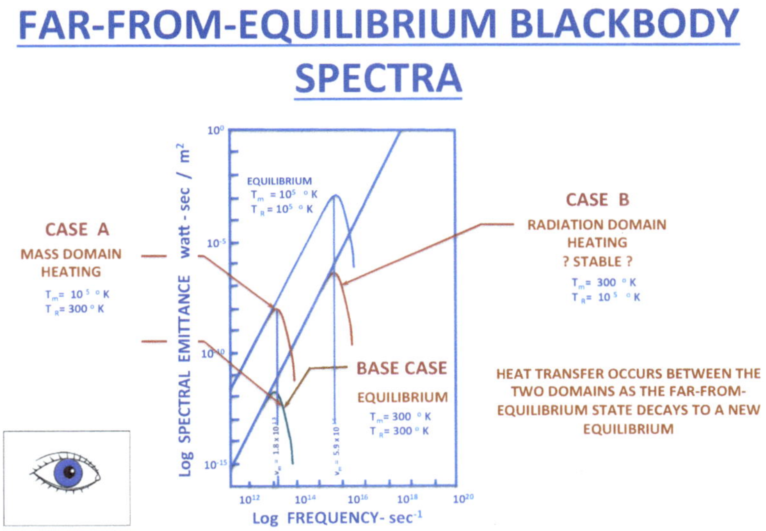

In living systems, it seems possible that these two temperatures could remain separated in a stable far-from-equilibrium configuration by masking a portion of the blackbody spectra within the mass system's covalent bond structure. In this way, the radiation energy is isolated in a stable manner that uncouples it from the normal mechanisms of energy redistribution, and prevents its return to equilibrium. By masking a portion of the energy in this way, large quantities of heat energy can be stored without increasing the system's apparent temperature.

A third feature of the far-from-equilibrium spectral distribution is its enormous, but always finite, energy storage capacity. The far-from-equilibrium region shown for Case B represents trillions of individual quanta, each of which could, in theory, be coded with one bit of information. We will propose for the moment (because this simplicity will aid in explaining some difficult concepts), that each of these represents the energy storage in a single covalent bond. The total energy at any frequency increment can be shown to be:

(4) $$E_\nu = K(\nu)\Delta\nu \cdot {c^2}/{\nu_1^2}$$

where $\Delta\upsilon$ is the frequency interval $\nu_1 \pm$ and the number of quanta at each frequency is found by dividing (4) by $h\nu_1$

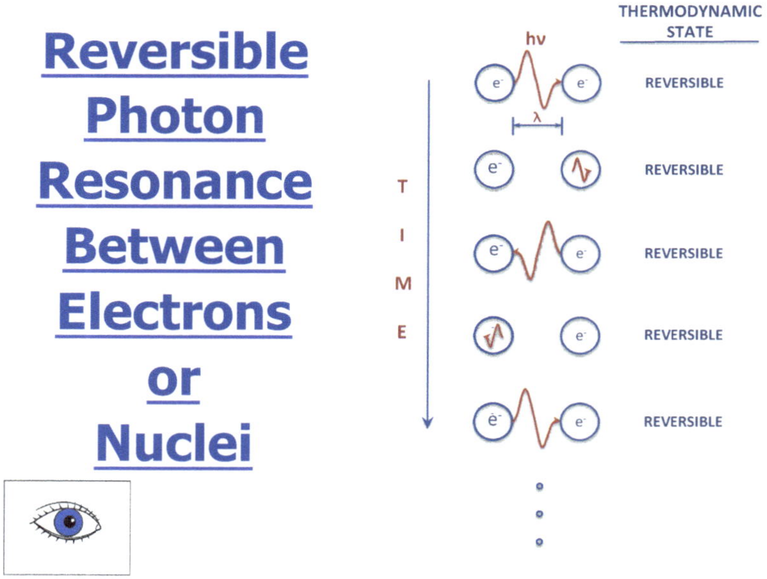

According to this theory, these covalent bonds exist as shared electro-magnetic quantities. The electro-magnetic waves are alternately absorbed and emitted, but always in accordance with the first step of the two-step absorption process. The light energy is completely absorbed in its one-dimensional form, but never decouples or transforms its electrical quantity to three-dimensions. Thus, it never undergoes the entropic transformation in the absorption process' second step. Instead, it is immediately re-emitted, and absorbed by its covalent electron, and so on. The electro-magnetic quantity never experiences dielectric loss. The covalent bond is said to be adiabatic invariant, and therefore, a completely reversible physical process.

If at this point, we inquire as to the amount of covalent energy storage that might occur within a cell, rough estimates are possible. Using a cell of 10^{-5}m diameter and 30% solids by volume, having an average covalent bond density of one per atom, I calculate $10^{12.6}$ covalent bonds. I have not yet calculated the number of quantum sites represented by the frequency intersects in Equation (3)'s far-from-equilibrium emittance spectra because T_R is, for the moment undefined, but a correspondingly large number that is entirely dependent on the separation between the thermodynamic and radiation temperatures.

From this theory's perspective, the number of available covalent bond sites at $293\,^oK$ could be the primary determinant of cell size. Furthermore, the relative invariance of cell size in the animal kingdom suggests that the separation between the mass and radiation temperatures is relatively constant across the spectra of living organisms. Warm-blooded species would appear to have the advantage of higher total energy storage for any given separation between the thermodynamic and radiation temperatures.

D. The Principle of Thermal Quiescence

The aqueous phase outside the cell membrane consists of a complex chemistry with one common characteristic, uniformity of thermal content. As we have already seen, this heat energy is partitioned into its two components: the radiation domain's blackbody spectra, and the mass domain's thermal motion. However, if we look closely at the way that the heat energy is partitioned between the two conditions, both sides of the membrane are very different.

If we first examine the aqueous milieu outside the cell membrane of an aerobic animal cell, we find small molecules that constitute the basic building blocks of cell metabolism (simple sugars and amino acids), as well as smaller ions and simple inorganic structures involved in gas (CO_2, O_2) and ion (H^+, NH_4^+, Ca^{++}, Cl^-, HCO_3^-, etc) exchange. Each of these exhibits a degree of thermal motion that is consistent with their size and temperature. This is even better illustrated at the exterior side of a plant's cell wall where the basic pre-transport structures are all very simple, and therefore, have significant thermal motion.

PRINCIPLE OF THERMAL QUIESENCE

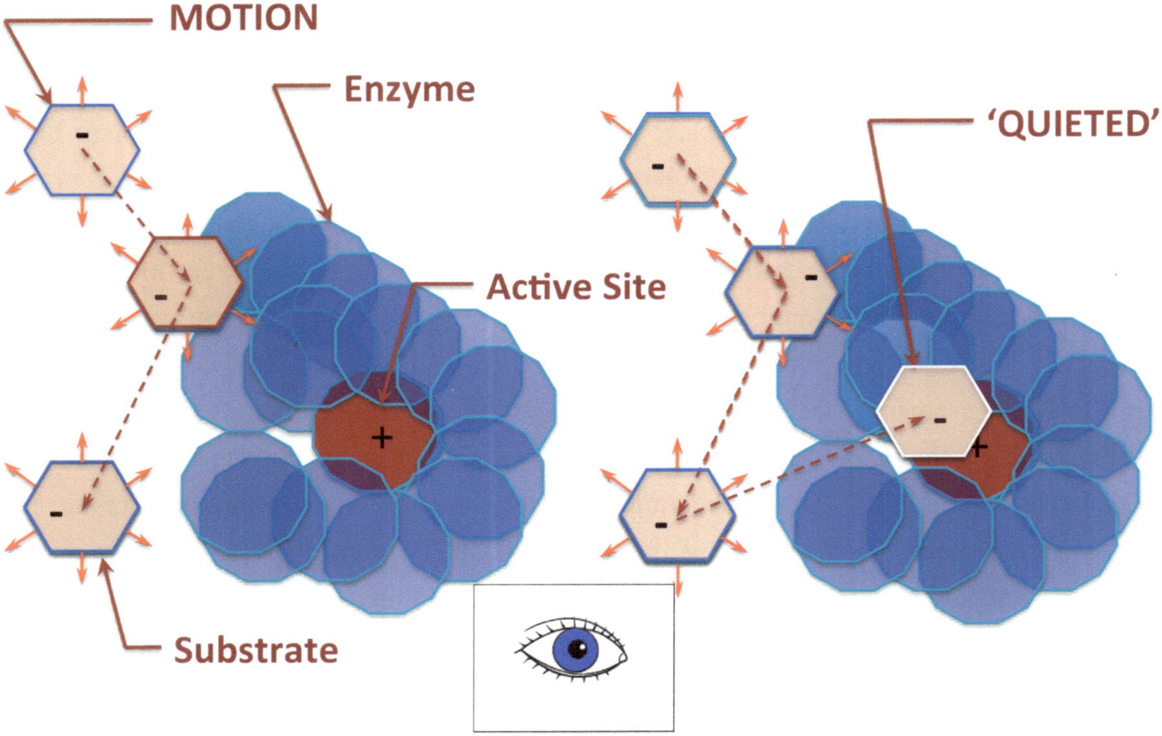

Within the animal cell's membrane or the plant's cell wall, the thermal condition is quite different. We find that these small sub-units, having been transported across the cell membrane, are metabolically transformed into progressively large molecules that have correspondingly little thermal motion. (In the case of the ions, many are simply transported outward again.) In effect the thermal motion of the precursor sub-units of cell synthesis, and the energy of motion that they initially contained, is no longer evident. It has been 'quieted'. The First Law requires that this energy be conserved. The Second Law tells us how this might occur, and also, something about its significance.

The operative thermodynamic system within the cell consists of cellular enzymes and their very specific substrates. The later might be the sub-units that were transported across the cell boundary, or enzyme mediated intermediates that in turn become substrates for higher order enzyme reactions. In this system, an enzyme protein cleaves its substrate(s) at its active site. This reduces the thermal motion of the substrate to essentially zero. It is important to note that the process of attachment at the active site is electrostatic, and completely reversible. (This condition can exist only if the surrounding medium is a near perfect dielectric, i.e. pure water.) If the electrostatic force is removed, the system reverts to its previous state, and vise versa. For the moment we will assume that there is no free energy loss associated with it.

This is not a statistical process where large numbers of molecules react in a reversible or irreversible manner. Indeed, some of the enzymes and intermediates exist at such low concentrations that only a few copies exist within a cell. In addition, the enzyme/substrate cleavage makes this chemistry fundamentally different than other classes of chemical processes. It takes place only at the microscopic scale where all events are reversible and the law of entropy increase has no meaning. This is the domain of Helmholtz, reversibility, and the Principle of Least Action.

According to the method of Helmholtz, the free energy, H, includes both the system's heat energy (including the operative chemical potentials) and its thermal motion. The relationship between the Helmholtz free energy and the systems total energy is known from thermodynamics:

(5) $$H = E - TS$$

Prigogine [10] refers to this equation as reflecting a competition between the system's energy, E, and its entropy, S.

In our system, there is no free energy loss, only a free energy partition, and that partition has only two parts: 1) the quieting of thermal motion by the enzyme, ξ, and 2) the substrate's decrease in chemical potential, μ, as they cleave. Accordingly, the Principle of Least Action may be written:

(5) $$d\left(\frac{\delta H}{\delta T}\right) = \mathrm{dS}$$

and as long as the thermodynamic temperature of the system is unchanged, the entropy decreases by an amount equivalent to the sum of ξ and μ, divided by the temperature:

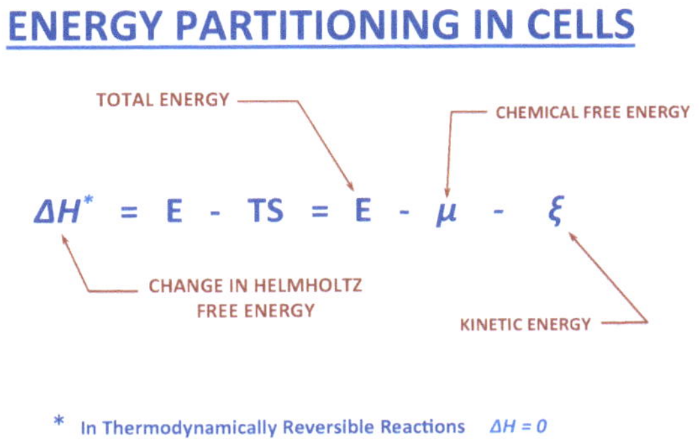

(6) $$S = -\frac{(\mu + \xi)}{T} = \frac{\mu}{T_R} + \frac{\xi}{T_m}$$

The net entropy decrease:

$$(7) \qquad S = -\frac{\xi}{T_m}$$

is the additional quantity that gives enzymes their ability to surmount energy barriers. In effect, the captured thermal motion's allows the enzyme to exploit an instability in the normal thermodynamics, causing a branch to what would otherwise be an inaccessible thermodynamic state. Nicolis and Prigogine [13] have called the ordered state that emerges in this manner a dissipative structure.

Following this reasoning further, using an amino acid chain as an example, we find that as the protein molecule increases in length, its thermal motion continually decreases; the terminal decrease occurring as the protein folds in on itself, forming its active condensed structure. Similar reductions in thermal motion occur throughout the cell interior. In all cases the biochemistry is reversible, and the thermal energy change shows up as an apparent entropy decrease. I call this principle, Thermal Quiescence.

Accordingly, the biochemical thermodynamics of Klotz [12] and Lehninger [13] might more appropriately substitute the Helmholtz free energy for that of Gibbs. In most instances, the difference tends to be small. However, substances such as molecular oxygen, a highly paramagnetic quantity, brings a higher negative free energy to the thermodynamic calculations, that favors reduction even more so than the Gibbs' thermodynamics indicate. Certain paramagnetic sulfur and iron molecules have similar characteristics, and also mediate respiratory metabolism.

ENZYME FUNCTION

Might exploit random thermal motion to form covalent bonds, and a localized decrease in entropy. In this way, the enzyme harvests waste heat of motion and converts it into free energy.

It is:

Adiabatic reversible process

Helmholtz free energy alone is operative

Principle of Least Action applies universally

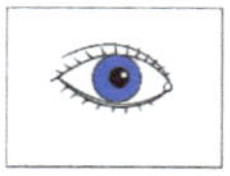

By way of summary, it appears that living systems, and in particular their enzyme activity, might exploit random thermal motion to effect covalent bond formation, and localized decreases in entropy. The process by which this occurs is an adiabatic reversible one, wherein the operative free energy measure is that described by Helmholtz. The cell is viewed as a localized region of reversible thermodynamic state, wherein the random motion of heat energy is harvested and transformed to chemical potential.

Electro-magnetic energy stored in the cells covalent bond structure causes small incremental increases in the systems radiation temperature alone, leaving the thermodynamic temperature of the cell unchanged. Thus, large amounts of energy accumulate in the cell without affecting its apparent temperature, and a stable far-from-equilibrium state occurs. The degree of separation in these two temperatures is a quantitative measure of that state.

E. The Cell's Covalent Bond Structure

The two temperatures, T_m and T_R, describe the total amount of energy that can be stored in the theoretical far-from-equilibrium spectral fabric. Separately, they represent the environment's thermal state (T_m), and what might be thought of as the intra-cellular potential (T_R). The later is a descriptor of the maximum number of covalent bond sites possible within a particular cell, at a particular time. It is this upper limit that is important in what follows.

The difficulty in using T_R is that it cannot be measured. A suitable analog might be found in redox potential, pe, and more specifically the electron activity, $\{e^-\}$ within the cell. pe is defined as $-log_{10}\{e^-\}$ in a way that is similar to pH ($-log_{10}[H^+]$. The cell is then described as a region of space having a high negative pe, and the electron rich cellular milieu is maintained by a continual influx of electrons across the cell membrane. These electrons are covalently bonded in enzyme mediated redox reactions, and contribute to small incremental increases in $\{e^-\}$, and T_R. It is this influx of electrons that is the forcing function for cell cycle changes in T_R.

For the moment we will not be concerned with the form of the forcing function that continually brings electrons across the cell membrane (Chapter 4). This will be taken as a given, and we will instead focus on the effects of electron influx.

The cell can be viewed as an enormous network built upon the energy structure of the far-from-equilibrium blackbody spectra. The last position filled was the lowest-energy spectral quantity that had been unfilled. Accordingly, there is in this spectral distribution one, and only one, position that is the next covalent bond. It awaits the addition of one more electron and has the distinction of being the covalent bond selected for by the Principle of Least Action. Regardless of where that electron is input to the system, the most likely location to which it will 'tunnel', is predetermined.

What then determines the molecular change that will occur within the cell? It is the change that occurred the last time that this cell's predecessors were in precisely the same redox state. Everything that has allowed this state to develop has followed an exact sequence. The predecessor molecules are in place. They may even be in exactly the same relative positions that the evolution of this state has predetermined. And although there may be an extraordinarily large number of things happening within the cell related to disease, environmental state, nutrition, etc., there is only one next step in the protein synthesis progression which is inexorably carrying the cell toward mitosis.

All of the processes that have been observed in the last 200 years of biological research continue to exist within this model. All that we know about the genetic predisposition in DNA, RNA, mRNA, etc., and protein synthesis, are still operative. But now there is a single process building the cell's biochemistry on the scaffold, which is the far-from-equilibrium blackbody spectrum. It is a linear progression from one redox state to the next that is unerring in its fidelity to what has come before.

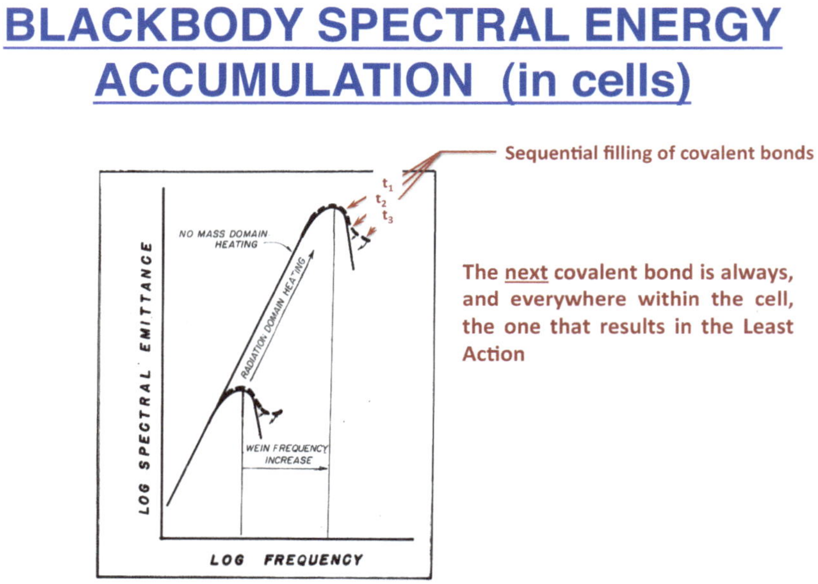

If for some reason, this progression is altered in the slightest way, modifications to the cell's protein structure occur, and at some subsequent step in that cell cycle there is a variant in how the next lowest energy level is filled. Although it is not clear how this occurs, differentiation of form, and perhaps function, occurs. This could be a mutation, or a branch to a different tissue type.

F. Cell Cycle Changes in the Covalent Structure

We will treat the cells DNA as a boundary condition. More precisely, the DNA's covalent bond structure might be thought of as filled with high energy sites that emit at the far right hand side of Curve B. These describe the maximum T_R in the cell's far-from-equilibrium state and a corresponding redox potential. The intervening frequency domain of a G_1, S and G_2 phase cell contains a region that cell synthesis fills with precisely determined numbers of covalent bonds as the redox potential increases in response to electron uptake. The ordering of this process is, for practical purposes, unerring in its precision.

CELL CYCLE ENERGY STRUCTURE AND CELL DIVISION CONSEQUENCES

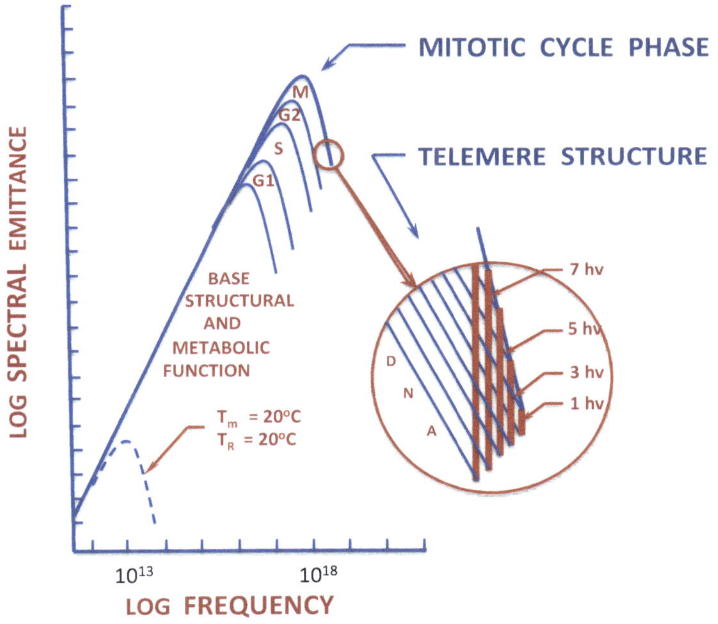

The illustration shows a generalized construct of the cell's electro-magnetic signature at several stages in the cell cycle. Curve M represents the way that the far-from-equilibrium spectral distribution is filled at, or very near to, cell division. Curve G_1 is the condition following cell division. Half of the molecular and heat radiation content in the dividing cell has gone to daughter cells, leaving a large unfilled region in the spectrum's energy structure, and also leaving the cell at a lower redox state. Memory of the filled condition exists, for example, in the cell's DNA. Region S represents what the spectral distribution might look like during the cell cycle's S stage. The region between the G_1 or S curves, and the M curve can be thought of as the cell's 'synthesis potential'. It is this covalent structure that must be synthesized before the cell can divide again.

The expanded section of the spectral curve shows how the covalent bonds are filled at the DNA boundary late in the cell cycle's S phase. The lowest diagonal line represents the spectral emittance of a single covalently localized quantum, hv. This is the highest energy state possible at the operative T_R during this mitotic cycle. However, this site is never replicated, and is lost at each cell division. In fact, the last filled site at each frequency in the spectra, is lost following cell division because fractional quantum states cannot exist. This effectively reduces T_R by 'quantum' amounts during each cell cycle, and thereby reduces the synthesis potential in the daughter cells.

Subsequent generations have slightly shorter DNA strands (telomeres) because the highest energy sites are the first to become vacant as T_R and $\{e^-\}$ drop. Eventually T_R drops to a level where the synthesis potential is too small to support replication. Furthermore, assuming that the cell's more basic functions, those associated with respiration and metabolism, are located in the lower energy portion of the spectra, their fractional reduction is smaller, and their function is only slightly effected, but never the less, diminished. This is the aging process.

G. Alterations to the Upper Boundary Condition

Now, let's assume that we change the upper boundary condition in plausible ways. Are the consequences equally plausible when compared to known cell behavior? For example, introduce into the cell a substance having a covalent bond structure that is higher energy than that in the cell's DNA (existing farther to the right). We will have to introduce this substance at sufficient concentration (or over a sufficiently long time) to significantly increase the radiation temperature within the cell, and thus, increase the upper boundary condition beyond that defined by the DNA itself. This adjustment is opposed to the aging process, which moves T_R to the left. Instead, it increases T_R, introduces useless information into the upper boundary condition, and causes excessive synthesis including extraneous DNA.

The introduced substance is a carcinogen. If its concentration is not large enough to change the DNA pool in a significant ways, its effect remains in the boundary condition, and further environmental exposure compounds the effect. This suggests why cancer is a disease of old age.

In a similar way, high-energy radiation exposure can alter the upper boundary condition by increasing T_R directly. This energy has to be captured within the cell as a covalent bond structure, and again, to the right of the DNA boundary condition. Once T_R is increased in a stable manner, regardless of its cause, or the final molecular states, synthesis potential increases.

Virus can have the same effect as long as its upper covalent bond energies are higher than those in the host cell's DNA. If the virus' covalent signature is within the synthesis region, the virus is simply multiplied.

In all of these cases, so long as the cell is still viable, the altered biochemistry and DNA will continue to replicate indefinitely, or at least until T_R, the synthesis potential, diminishes to the levels where replication ceases. In this way, aging and cell proliferation both appear in this theory as changes in the cell's DNA (electro-magnetic) boundary condition, or more to the point the redox state defined by that boundary condition.

AGING AND CANCER ARE ALTERATIONS TO THE UPPER BOUNDARY CONDITION

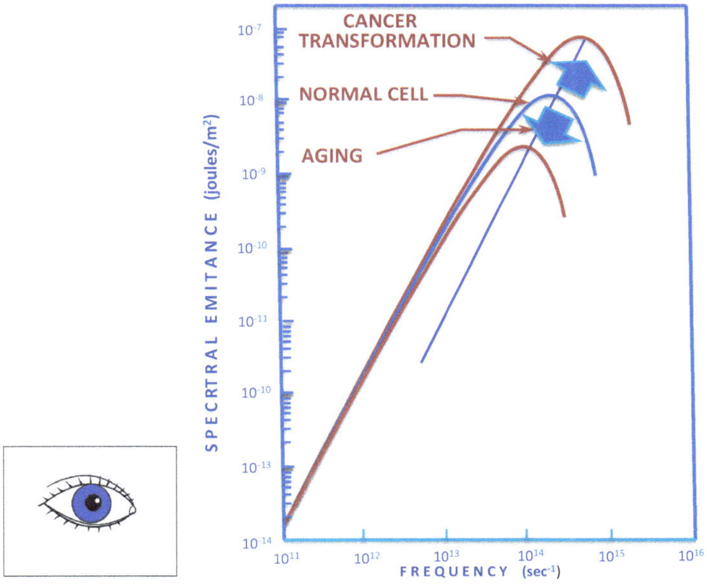

Aging is the condition where high-energy covalent bonds are systematically eliminated and the maximum attainable redox potential is reduced below the minimum required for mitosis. Cancer is a condition where the redox state is increased to levels exceeding that required for mitosis, and the cell machinery continues duplicating extraneous, high energy, covalent bonds until mitosis occurs…again and again and again.

H. Discussion

This theory is speculative. But then, what theory of the living state is not. Its speculative nature should not diminish the underlying concepts that brought us to this point. The first of these is our continuing emphasis on the molecular quantities, when it is entirely possible, and even probable, that the electro-magnetic signature within the cell will be at least as relevant, and provide new insight into what we now regard as life's secrets. The radiation domain is also simpler. Only three or four degrees of freedom appear to be operative: frequency and two temperatures.

Secondly, the cell's negentropic character implies reversible thermodynamic behavior according to the methods of Helmholtz. This removes the mystery from enzyme reactions, and permits us to see life for what it is: a process that harvests the random heat of motion, and transforms it into electro-magnetic, covalent bond energy, and localized decrease in entropy. Here, the second law operates at its lower limit: a zero net change in the cell's free energy state.

Third, the far-from-equilibrium blackbody spectrum seems ideally suited for understanding far-from-equilibrium systems such as the living cell. It is implicitly a time dependent operator with time vectors in opposing directions, $\left({}^{t_2}/_{t_1}\right)$. One is the entropic direction that we live in, while the other is a negative time vector that modifies time's passage by slowing it down. In the limit, the local time stops, and physical processes enter the realm of reversibility, and the Principle of Least Action prevails. The function $f_3(t_2/t_1)$ (where $t = {}^{1}/_{\nu}$) can be thought of as a true entropy operator.

THE CELL'S NEG-ENTROPIC CHARACTER...

....IMPLIES REVERSIBLE THERMODYNAMIC BEHAVIOR

Studies of the cell's joint energy-entropy structure, particularly those focused on its universal character, delineate areas that are ripe for study.

The literature is rich with research documenting: redox correlation with the cell cycle [14-17], cell cycle regulation [18,], redox regulated biochemistry [19-28], redox states in different cell organelles [29], Redox control in plant metabolism [30,31], and redox branching points in development [32-34], aging [35,36], death [37-40], and cancer [41-43]. This paper's proposal is a modeling framework upon which these observations can be organized in their relationship to one another. But, it goes further. In understanding the hierarchy for redox state progression over a cell cycle, we stand poised to mathematically model normal and aberrant cell function. The one essential feature of a coherent cell modeling framework that is not included in the preceding, is a mechanistic and energetic description of the forcing function(s) responsible for the continuous influx of electrons to the cell. This does not diminish this work's form and function arguments. It does, however, leave an unsatisfying gap that will have to be dealt with separately in Chapter 4.

One final thought …I have elected to place the energy quanta associated with the far-from-equilibrium blackbody form into the cell's covalent bond structure. This is not only reasonable, but also convenient because it is simple and understandable. However, this is not the only possibility, and indeed, there is a growing body of evidence that the energy storage could also be localized in excited nuclear states. In particular, researchers beginning in the 1950's have assembled scientific data suggesting that nuclear transmutations are occurring in biological systems. If that turns out to be the case, it would be difficult to achieve T_R's approaching solar core temperatures with covalent bond energy alone. The only way that I am aware of for storing such extremely high energy, in a thermodynamically reversible manner, is in excited nuclear states that are in Mossbauer resonance. This places high-energy quanta resonating between identical nuclei in a completely reversible state similar to that in the covalent bond, but at energies in the realm of gamma quanta.

FAR-FROM-EQUILIBRIUM ENERGY STORAGE

- COVALENT BONDS BETWEEN ELECTRONS
- MOSSBAUER RESONACE BETWEEN NUCLEI

I. Conclusions

The far-from-equilibrium blackbody equation that lies at the core of this study's cell model describes a reasonable framework for understanding cell form and function. The blackbody form has long been known to represent the spectral distribution of energy in inanimate matter. The choice of a non-equilibrium form of the equation as a framework for energy storage in matter's living state is logical, and appears to be capable of providing useful insights.

It is also appropriate that the theory presented here distinguishes between the thermodynamic temperature and the radiation temperature. The thermodynamics presented in Section D show how the cell harvests heat of thermal motion from its environment and accumulates it within the cell's covalent bond structure. In essence, the harvested heat 'raises the temperature' at the interior of the cell, but this increase is not apparent because the electro-magnetic radiation is all masked within the cell's covalent bond structure. T_R is a measure of the cell's negentropy content. Burning the cell liberates the heat equivalent of this negentropy.

This paper's findings, although speculative, appear to provide some meaningful insights into cell processes. It is particularly noteworthy that an internally consistent interpretation provides insights to diverse cell features, including: cell size, aging and cell proliferation. It also appears that advances in the cell's redox state drive both synthesis and metabolism forward in time.

The theory also suggests that protein synthesis pathways are very exact progressions that retrace those of the cell's antecedents…progeny recapitulate endogamy. However, this does not allow for the kind of specificity that occurs in differing tissue types within a single organism. It would appear that there is a set of synthesis pathways that can be altered by the cell's local environment within the organism, or perhaps, there are more specific synthesis sequences that include the very exacting branches that tissue differentiation requires. The former seems more plausible to this researcher.

Finally, we should note that Maxwell's demon [44,45] is present in this cell model in a very real way. Here it assumes an enzyme identity, and as in Maxwell's example, selectively harvests heat from its environment, and converts that environment to a more ordered state. This circumstance does not allow us to draw on the information processing ability of intellect to dismiss the demon's magic as Szilard [46] did. However, it is clear that a vast amount of negentropy is concentrated in the enzyme structure, information that can be drawn on to quiet substrate molecules, and order them accordingly.

NEXT CHAPTER PREVIEW:

THE CELL'S REDOX STATE DRIVES SYNTHESIS AND METABOLISM FORWARD

The ideas presented here are an evolving process directed at understanding the nature of the living state. This is but another approach that is brought to the table for debate and criticism. As such it is not an answer in itself.

J. References

(1) Elsasser, W.M., Reflections on a Theory of Organism, Johns Hopkins University Press, Baltimore, 1987.

(2) Schrodinger, E., What is Life?, The Physical Aspects of the Living Cell, MacMillan Co, NY, 1946.

(3) Szent-Gyorgyi, A., The Study of Energy levels in Biochemistry, Nature148: 157-159, 1941.

(4) Monod, J., Chance and Necessity, Alfred A. Knopf, NY, 1971.

(5) Luria, S.E., Life - The Unfinished Experiment, Charles Scribner's Sons, NY, 1973.

(6) M. Planck, Verhandlunger der Deutschen Physikalischen Gesellschaft, 2, 237, (1900), or in English translation: in H. Kangro (ed.), Planck's Original Papers in Quantum Physics, Volume 1 of Classic Papers in Quantum Physics, Wiley, New York, 1972.

(7) Boltzmann, L., Lectures on Gas Theory, Translated by Stephen G Brush, University of California Press, Berkeley, 1964.

(8) Maxwell, J.C., A Treatise on Electricity and Magnetism, Clarendon Press, Oxford, 1892.

(9) Szumski, D.S., Theory of Heat I - Non-equilibrium Blackbody Radiation Equation, unpublished manuscript, 2000.

(10) Prigogine, I., From Being to Beginning, W.H. Freeman and Company, San Francisco, 1980.

(11) Nicolis G., Prigogine I., Self-Organization in Nonequilibrium Systems, John Wiley and Sons, NY, 1977.

(12) Klotz, I., Energy Changes in Biochemical Reactions, Academic Press, NY, 1967.

(13) Lehninger, A., W A Benjamin, Inc., NY, 1965.

(14) Yukihashi, Y, Yashumitsu, O, Kazuo S, Synchronized generation of reactive oxygen species with the cell cycle, Life Science, 75: 301-311, 2004.

(15) Conour, J, Graham, W, Gaskins, H, A Combined In-vitro/bioinformatics Investigation of Redox Regulatory Mechanisms Governing Cell Cycle Progression, Physiol. Genomics, 18: 196-205, 2004.

(16) Boonstra, J, Post, J A, Molecular events associated with reactive oxygen species and cell cycle progression in mammalian cells. Gene 337: 1-13, 2003.

(17) Menon, S G, Goswami, P C, A redox cycle within the cell cycle: Ring in the old with the new, Oncogene, 26: 1101-1109, 2007.

(18) Schackelford, R, Kaufmann, W, Paules, R, Oxidative stress and cell cycle checkpoint function, Free Radical Biology and Medicine, 28: 1387-1404, 2000.

(19) Nakamura, H, Nakamura, K, Yodoi, J, Redox Regulation of Cellular Activation, Annu. Rev. Immunol.,15:351-69, 1997.

(20) Ramos, K, Redox regulation of c-Ha-ras and osteopontin signaling in vascular smooth muscle cells: implications in chemical atherogenesis, Annu. Rev. Pharmacol. Toxicol, 39:243-265, 1999.

(21) Irani, K, Oxidant signaling in vascular cell growth, death, and survival: A review of the roles of reactive oxygen species in smooth muscle and endothelial cell mitogenic and apoptotic signaling, Circ. Res., 87: 179-183, 2000.

(22) Fernandes, A P, Holmgren, A, Glutaredoxins: glutathione-dependent redox enzymes with functions far beyond a simple thioredoxin backup system, Antioxidants & Redox Signaling, 6: 63-74, 2004.

(23) Sun, Y, Oberley, L W, Redox regulation of transcriptional activators, Free Radical Biology & Medicine, 21: 335-348, 1996.

(24) Sen, C K, Packer, L., Antioxidant and redox regulation of gene transcription, The FASEB Journal, 10: 709-720, 1996.

(25) Rhee, S G, Chang, T, Bae, Y S, Lee, S, Kang, S W, Cell regulation by hydrogen peroxide, J Am. Soc. Nephrol., 14: S211-S215, 2003.

(26) Rhee, S G, Redox signaling: Hydrogen peroxide as intracellular messenger, Experimental and Molecular Medicine, 31: 53-59, 1999.

(27) Powis, G, Briehl, M, Oblong, J, Redox signaling and the control of cell growth and death, Pharmac. Ther., 68: 149-173, 1995.

(28) Herrlich, P, Bohmer, F D, Redox regulation of signal transduction in mammalian cells, Biochemical Pharmocology, 59: 35-41, 1999.

(29) Schafer, F Q, Buettner, G R, Redox environment of the cell as viewed through the redox state of the gluathione disulfide/glutathione couple, Free Radical Biology & Medicine, 30:1191-1212, 2001.

(30) Buchanan, B, Balmer, Y, Redox regulation: A broadening horizon, Annu. Rev. Plant Biol., 56: 187-220, 2005.

(31) den Boer, B, Murphy, J, Triggering the cell cycle in plants, Trends in Cell Biology, 10: 245-250, 2000.

(32) Dennery, P, Role of redox in fetal development and neonatal diseases, Antioxidants & Redox Signaling, 6: 147-153, 2004.

(33) Smith, J, Ladi, E, Mayer-Proschel, M, Noble, M, Redox state is a central modulator of the balance between self-renewal and differentiation in a dividing glial precursor cell, Proc. Natl. Acad. Sci, 97: 10032-10037, 2000.

(34) Harvey, A, Kind, K, Thompson, J, Redox regulation of early embryo development, Reproduction, 123: 479-486, 2002.

(35) Hagen, T, Oxidative stress, redox imbalance, and the aging process, Antioxidants & Redox Signaling, 5: 503-506, 2003.

(36) Lavrovsky, Y, Chatterjee, B, Clark, R A, Roy, A K, Role of redox regulated transcription factors in inflammation, aging, and age related disease, Experimental Gerontology, 35: 521-532, 2000.

(37) Ueda, S, Masutani, H, Nakamura, H, Tanaka, T, Ueno, M, Yodoi, J, Redox control of cell death, Antioxidants & Redox Signaling, 4: 405-414, 2002.

(38) Kwon, Y, Masutani, H, Nakamura, H, Ishil, Y, Yodoi, J, Redox Regulation of cell growth and death, Biol. Chem., 384:991-996, 2003.

(39) Hall, A G, The role of glutathione in the regulation of aoptosis, European Journal of Clinical Investigations, 29: 238-245, 1999.

(40) Nakashima, I, Suzuki, H, Kato, M, Akhand, A, Redox control of t-cell death, Antioxidants & Redox Signaling, 4: 353- 356, 2002.

(41) Loo, G, Redox sensitive mechanisms of photochemical-mediated inhibition of cancer cell proliferation, Journal of Nutritional Biochemistry, 14: 64-73, 2003.

(42) Hoffman, A, Spetner, L M, Burke, M, Cessation of cell proliferation by adjustment of cell redox potential, J. Theor. Biol., 211: 403-407, 2001.

(43) Aw, T Y, Cellular redox: A modulator of intestinal epithelial cell proliferation, News Physiol. Sci., 18: 201-204, 2003.

(44) Maxwell, J.C., Theory of Heat, reprinted Dover, New York, 1871.

(45) Brillouin, L., Maxwell's Demon Cannot Operate: Information and Entropy, J. of Applied Physics, 22, 334-337, 1951.

(46) Szilard, L., On the Decrease of Entropy in a Thermodynamic System by the Intervention of Intelligent Beings, translated by Anatol Rapoport and Cechilde Knoller, in B.T. Feld and G.W. Szilard(ed), The Collected Works of Leo Szilard- Scientific Papers, MIT Press, 1972.

Chapter 4

Understanding Determinism and Variability in Living Cells

A. Introduction

My previous paper (1) on this subject introduced a mathematical modeling framework for understanding cell structure and function. Its focus was on the deterministic portion of cellular biochemistry, as seen through the lens of the cell's overall energy structure. Here, the focus is very different. We will be studying the mechanism that introduces systematic variability into cellular processes. This mechanism produces what is seen as a high degree of stochastic response into our data, even though as we will see in what follows, it is entirely deterministic. Our goal here is understanding the source of apparently random variations that confound our ability to construct precise mathematical models of biological processes, like those that we see in physics and chemistry. This strikes at the very core of the conundrum that impedes our progress toward the ultimate goal of quantitative biology.

One might ask: "How can we account for variability in biological response, and do so in a way that is itself deterministic?" In this paper we will use methods from mathematical physics that will uniquely determine that next step in the seemingly stochastic process that produces variability in biological system response. We will do this with the precision of physics. Our results will be exact, and reproducible.

LIFE IS...

...A STABLE, LOCALIZED REGION OF MATTER'S FAR-FROM-EQUILIBRIUM STATE

My purpose in preparing the four manuscript in this series, is founded in my belief that the process that gives rise to matter's living state is not understood at its fundamentals because of an over emphasis on the cell's mass quantities… the things that we can see and measure in living cells. This arrangement of mass is fundamental to our most basic definition of life: a stable, localized region of matter's far-from-equilibrium state.

What I believe is missing from our understanding, is the nuance introduced into biology by considering the unseen portion of the cell; the energy structure that 'glues' everything together. Surely that energy structure must be every bit as far-from-equilibrium as the cellular structures that we can see. Indeed, if we want to be dramatic in making this point we might say that the mass quantities hold one half of life's secret, while the other half is tied up in the cells internal energy structure. Let's give a name to this unseen part of the cell. Covalent bond structure might be a good first approximation. It is something that cannot be seen, but we know that it exists. And, for our purposes, it will be characterized by its complexity, and its permanence.

Try to imagine the complexity of this covalent bond structure. But, let's be clear. We are not talking about individual covalent bond energies. Rather, we are looking for the overriding mathematical structure that defines the energy intensity of all covalent bonds within a cell. It might be visualized as a spatial network of the individual covalent bond energies, or as it turns out, their organization according to the frequencies of the individual energy quanta within the covalent bonds. It is likely that this energy network is as far-from-equilibrium as the state of matter found in the cell's molecular associations.

Permanence is seen in a piece of wood recovered from an Egyptian tomb. The wooden sample's cellular structure remains unchanged for centuries. The covalent energy content of the wood sample can be measured if we burn it. But more to the point, the energy content from burning that wood sample would be the same now, as it will be if we were to burn it one thousand years into the future. In other words, there is permanence to the energy storage in the piece of wood that does not in any way diminish in time. We say that the process that holds that energy in the wood sample is in accordance with the laws of reversible thermodynamics. A similar argument can be made if we instead consider mitochondrial DNA from the tissue sample of an extinct species or a chunk of coal. By contrast, all irreversible thermodynamic processes involving electrical energy experience a systematic diminution of electrical field strength into heat of molecular motion due to dielectric loss. Now let's see if we can figure out how this reversible process exists in the living state that the wood or DNA sample was derived from.

B. The Current Paradigm in Biology is not Sufficient

We have begun to lay the groundwork for addressing the biological conundrum: biology differs from other physical sciences in that its theories are not mathematical. This stems from the belief that biological process are highly variable, and therefore not amenable to the type of deterministic treatment that is common in mechanics, electricity and magnetism, optics, astronomy, or aqueous chemistry. However, there is enough of a deterministic response that precise biochemical pathways can be described, as can the overall flow of cell synthesis and cellular metabolism. But, what is generally missing is model precision.

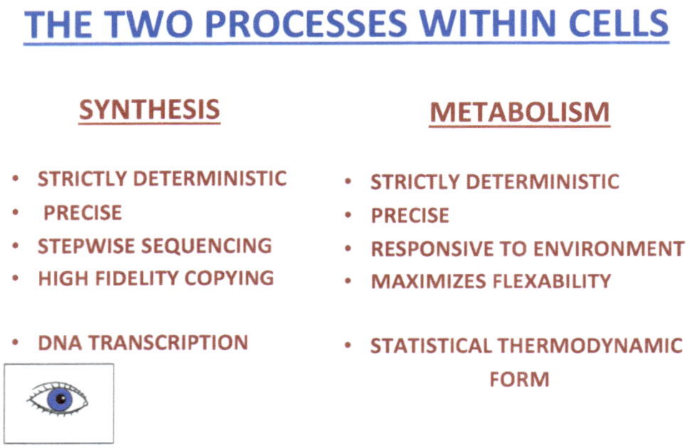

If we want to come to the truth of this matter, we need to objectively disaggregate our study of biology into two very fundamental components: synthesis and metabolism. I will now show how the first of these is a very precise process that is unerring in its temporal evolution, while the second permits the broadest latitude of systemized response to the highly varied environmental circumstances that the living cell might encounter.

Synthesis is the process by which cells replicate. It is the very precise process stated in biology's Central Dogma (2), and that reads instructions from the DNA template and then proceeds through a detailed step-by-step process to assemble an exact copy of the parent cell. The unerring accuracy of this stepwise process is seen in the high fidelity copying of specific proteins, or more broadly in the high fidelity replication of uni-cellular and multicellular organisms.

Metabolism, on the other hand is the cell's link with the real world. It can be influenced in predictable, and seemingly unpredictable ways by varying environmental factors such as temperature, chemical makeup of the environment, nutrition availability, or disease factors. In other words the flexibility of metabolic pathways insures to the extent possible that the cell will remain viable through a broad range of environmental circumstances. In looking at the cell through this filter, we can see that its high degree of variability results from metabolic pathways rather than from the synthesis pathway (singular). Thus, the first rule of biology:

Rule 1: At its most basic division, cell function can be divided into synthesis and metabolism. Synthesis is the very precise sequence of steps that results in cell division and identical daughter cells. Metabolism is the cell's interface with an ever changing environment, and possesses all of the myriad pathways to accommodate change.

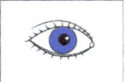

The objective of quantitative biology then becomes the mathematizing of mechanistic pathways for cell synthesis and metabolism. As many of us have discovered, this is no simple task. For example, experiments in metabolism yield plausible explanations of Monod kinetic behavior in enzyme reactions, and also the systematic behavior in O_2/CO_2 uptake and release by hemoglobin according to the Bohr effect. In both cases, the metabolic response and its rate are known to be influenced by several factors. We can vary those factors in experimental designs and arrive at a mathematical form for both processes. But this is not good enough for quantitative biology. We have to explore these and other processes to uncover the more fundamental law governing the process that both of these sub-processes have in common. In this way we arrive at what we might call the principle components of biology; a set of rules which all biology obeys in producing the specific responses that experimentalist observe.

In essence, we are trying to describe the most fundamental laws governing the structure and function of the cell. This is the skeletal structure of biology that all other process can be fitted into. Monod (3) called this the *ultima ratio*. Its equivalent in electromagnetism is found in Maxwell's six equations (4), in dilute chemical systems it is found in Gibbs theory (5), in mechanics, in Newton's equations (6) of motion. It is quite simply, the smallest set of rules that all of biology adheres to. And because biological systems evolve in space, time and mass/energy/entropy these are the parameters that we might expect in this all encompassing theory of the living state.

C. Respiratory Metabolism as a Reversible Process

We will begin by re-stating our fundamental definition of the cell. Instead of it being a localized region of matter's far-from-equilibrium state, we will introduce energy and entropy into our definition: the cell is a localized region of thermodynamically reversible processes. This immediately places several constraints on the cell model that we are developing. First, we must exclude all free electrical charges from our cell. Electrical charge is an entropic quantity that continually experiences dielectric loss. And, entropy production is something that is foreign to reversible thermodynamics. More simply stated we might say that electrical charge is always, and everywhere going over into the random heat of motion. Thus, it results in a process that is irreversible, and outside our definition of cellular processes.

THE CELL IS...

... a localized region of thermodynamically reversible processes

We can exclude electrical charge from our model in either of two ways: eliminating all ionic quantities $[H^+, Zn^{+2}, Na^+, Cl^-]$, or free electrons $[e^-]$, or any organic ion $[C_nH_mO_l^-]$. This is not a very practical way to proceed. However, the alternative is very attractive. In particular, we will require that the solvent in our cell has an extremely high dielectric constant, so that although charges might exist, they remain isolated, and conservative. Pure water satisfies this criterion (ϵ_{H_2O}= 78.54.). So does pure H_2O_2 ($\epsilon_{H_2O_2}$= 84.2).

Rule 2: Cellular water functions as a high strength dielectric that allows isolated electrical charge, and its attractive and repulsive forces, but without dielectric loss.

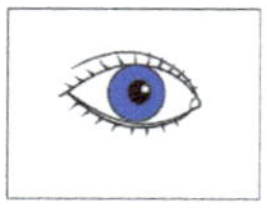

But how do we insure that the water in our cell is 'pure'? Again there are two ways to do this. The first is to make the cell membrane impermeable to ions that would otherwise partition into the cell water. This has the advantages of excluding charged quantities from the cell, something that we already discussed. However, the alternative is equally effective, but as it turns out, has some other distinct advantages that make it far more attractive. In particular, we are going to bring the water across the cell membrane by a catalyzed reaction:

(1) $$2[H]^{+} + O_2 + 2\{e^{-}\} \longrightarrow H_2O_2 \ ; \ E_v = 0.68\ volts$$

which is energetically coupled to one or more reactions of the enzyme (E) mediated form:

(2a) $$\left[E^{-2}\right] + 2\left[CO_2\right] + 2\left[H_2O_2\right] \xleftrightarrow{CA} \left[EH_2\right] + 2\left[HCO_3^{-}\right] + \left[O_2\right]$$

(2b) $$\left[Ca^{+2}\right] + 2\left[Cl^{-}\right] + 2\left[H_2O_2\right] \xleftrightarrow{CA} \left[Ca(OH)_2\right] + 2\left[HCl\right] + \left[O_2\right]$$

(2c) $$\left[EH_2\right] \xleftrightarrow{K^{\copyright}} \left[E^{-2}\right] + 2\left[H^{+}\right]$$

I will call this Reaction Sequence A. For the moment, and for illustration purposes, we will assume that the enzyme catalyzing these reactions is Carbonic Anhydrase (7), which cycles between its active and its fully hydrated form. Combining these three equations yield the very satisfying result:

(2d) $$2\left[CO_2\right] + 4\left[H_2O_2\right] \xleftrightarrow{CA} 2\left[HCO_3^{-}\right] + 2\left[H^{+}\right] + 2\left[H_2O\right] + 2\left[O_2\right] + \Delta F_f^{o}$$

$$(\Delta F_f^{o} = 18.46 \ \ kcal/mole)$$

This reaction links carbon dioxide excretion with oxygen uptake. Charged ionic quantities migrate toward the exterior side of the cell membrane, while neutral molecular quantities are partitioned into the cell.

Rule 3: All free energy exchanges within the cell are thermodynamically reversible, and according to the methods of Helmholtz.

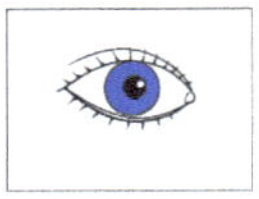

A second reaction sequence like the one above, advances this path of reasoning:

(3a) $$\left[E^{-2}\right]+2\left[CO_2\right]+2\left[NH_3\right]+2\left[H_2O_2\right]\xleftrightarrow{CA}\left[EH_2\right]+2\left[CO_3^-\right]+2\left[NH_4^+\right]+\left[O_2\right]$$

(3b) $$2\left[NH_3\right]+2\left[CO_3^{-2}\right]+2\left[Na^+\right]+2\left[Cl^-\right]+2\left[H_2O_2\right]\longleftrightarrow$$

$$2\left[HCO_3^-\right]+2\left[H_2O\right]+2\left[O_2\right]+2\left[NH_4^+\right]+2\left[Na^+\right]+2\left[Cl^-\right]$$

(3c) $$\left[EH_2\right]\xleftrightarrow{K^{©}}\left[E^{-2}\right]+2\left[H^+\right]$$

Adding these equations results in the overall Reaction Sequence B:

(3d) $$2\left[CO_2\right]+2\left[NH_3\right]+5\left[H_2O_2\right]\longleftrightarrow 2\left[HCO_3^-\right]+2\left[H^+\right]+2\left[H_2O\right]+3\left[O_2\right]+2\left[NH_4^+\right]+\Delta F_f^o$$

$$(\Delta F_f^o = -23.54 \quad kcal/mole)$$

But let's not stop here. Let's add one more reaction sequence to this grand ensemble. Notice that Sequence A transports C, H and O across the cell membrane. Sequence B then adds nitrogen transport. How about adding a third reaction Sequence C that includes phosphorus transport in an ATP generating step:

(4a) $$\left[E^{-2}\right]+2\left[NH_2COOH\right]+\tfrac{3}{2}\left[H_2O_2\right]\longleftrightarrow\left[EH_2\right]+2\left[CO_3^{-2}\right]+2\left[NH_4^+\right]+\left[O_2\right]$$

(4b) $$\left[H_2PO_4^-\right]+\left[Fe^{+3}\right]+2\left[CO_3^{-2}\right]+\tfrac{7}{2}\left[H_2O_2\right]\longleftrightarrow$$

$$2\left[HCO_3^-\right]+3\left[H_2O\right]+2\left[O_2\right]+2\left[NH_4^+\right]+\left[Fe^{+3}\right]+2\left[H^+\right]+ATP$$

(4c) $$\left[EH_2\right]\xleftrightarrow{K^{©}}\left[E^{-2}\right]+2\left[H^+\right]$$

The combined reaction for Reaction Sequence C yields one molecule of ATP and can be summarized:

(4d) $$\left[H_2PO_4^-\right]+2\left[NH_2COOH\right]+5\left[H_2O_2\right]\longleftrightarrow 2\left[HCO_3^-\right]+2\left[H^+\right]+3\left[H_2O\right]+2\left[NH_4^+\right]$$

$$+3\left[O_2\right]+\left[ATP^{-4}\right]+\Delta F_f^o, \qquad (\Delta F_f^o = -231.49 \quad kcal/mole)$$

This third reaction sequence utilizes -231.49 kcal/mole and forms one mole of ATP as part of the respiratory exchange of gasses. Ammonium carbomate, a respiratory product that expels additional quantities of carbon dioxide and ammonia, occurs in this sequence as a backup mechanism for expelling these toxic metabolic products.

Reversible reactions (8) always require the free energy treatment of Helmholtz, and never that of Gibbs. The above free energy utilization and production are given as Helmholtz free energy changes. This treatment includes in the free energy of reaction, that associated with latent molecular motion in the system, something that Gibbs free energy does not include.

We have also achieved our objective of forming pure water from its constituents: H^+ ions, molecular O_2, and free electrons orginating in the cell's nutrition source (animal cells) or photon absorbtion (plant cells).

Look at what else we have accomplished. In the first set of equations, we have linked the respiratory gas exchange of CO_2 and O_2 with a chloride shift. We have done this in an exo-thermal reaction having a free energy change of +18.46kcal/mole. We will call this Reaction Sequence A. It can occur spontaneously as indicated by its positive net free energy change. However, if this were to occur, the reaction would have net energy production, and thus, not satisfy Rule 3.

The second set of equations completes the gas exchange by including NH_3 gas and adding sodium to the chloride shift. This second set of coupled reactions is endo-thermal requiring -23.54 kcal/mole. We will call it Reaction Sequence B. Finally; Reaction Sequence C makes obligatory ATP formation a part of cellular respiration at an energy cost of -231.49 kcal/mole. These last two reaction sequences cannot occur spontaneously because of their energy requirements.

Rule 4: Metabolic stasis is governed by the rules of statistical thermodynamics, and thereby regulated with the broadest range of compensations for environmental change.

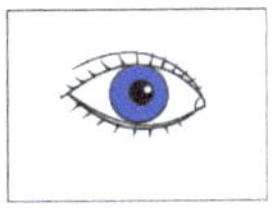

D. Metabolic Variability

I developed these reaction sequences many years ago while studying respiratory gas exchange at the gill epithelial membrane of teleost fish (9,10). That study attempted to describe the ammonia toxicity effect. Nevertheless, these reaction sequences serve an important purpose here. They illustrate how the broad range of influence factors affecting the respiratory exchange of gases might be linked mechanistically. My purpose here is not to create new science, but rather to demonstrate how a plausible mechanistic framework can unite in a single equation, all of the factors implicated in the variability of respiration.

We are now faced with the task of reconciling our observation that Reaction Sequences B and C have a net negative free energy yield, and consequently are not spontaneous. This is reconciled if we hypothesize a hybrid reaction combining A with the net negative free energy from reaction sequences B and C. There is an easy way of accomplishing this. Our initial assumption that the cellular process is thermodynamically reversible requires that the overall reaction has a net zero free energy change.

(5) $$18.46 \cdot A - 23.54 \cdot B - 231.49 \cdot C = 0.0 kcal/mole$$

where $A, B,$ and C refer to the number of reactions of each type in this reversible process.

This equation expresses the dynamic equilibrium condition necessary for all three reaction sequences to proceed in accordance with the reversible thermodynamic of Helmholtz. Any change in the stoichiometry (number of reaction sequences B or C), of necessity blocks the entire hybrid reaction sequence, or more specifically blocks Reaction Sequence A, the primary CO_2/O_2 exchange mechanism in this theoretical argument.

Electro neutrality requires that Reaction Sequences B and C occur in a 1:1 relationship, Thus: $B = C$, and the solution to Equation (5) is:

(6) $$A:B:C = 13.28:1:1 = 14:1:1$$

We can now get some idea of how metabolic flexibility is achieved in the cell. We begin by multiplying Reaction Sequence A by 14, adding one each of Reaction Sequences B and C to it, and viewing the overall reaction as a statistical thermodynamic process where the denominator includes the reactants, and the numerator the products, and the net energy change is identically zero:

$$(7) \qquad \frac{[HCO_3^-]^{32}[H_2O]^{33}[NH_4^+]^4[ATP]^1}{[CO_2]^{30}[H^+]^{96}[O_2]^{60}\{e^-\}^{128}[NH_2COOH]^2[NH_3]^2[H_2PO_4^-]^1} = e^{-\Delta F_f / k_b T} = 1.0$$

This may only be a partial representation of the overall respiratory metabolic stasis. But it does suggest the type of compensations that the cell is capable of achieving in response to environmental changes. This ratio, which I will call the cells 'metabolic stasis', varies in response to environmental changes, such that the delta free energy is always exactly zero. This is not a precise process yet. It has many degrees of freedom. However, it is very exact. We will await the introduction of the Principle of Least Action to make our prediction precise.

When I say that the process is exact, I am noting how, if any influence factors in the denominator changes, the products of respiratory exchange in the numerator compensate, and vice versa, but always under the zero net free energy change constraint. In this way, cellular processes remain entirely within the domain of reversible thermodynamics, and exactly regulated. Temperature effects are implicit.

However, there is a more important conclusion here that strikes at the very core of the problem that we had set out to solve. In particular, we see that it is this metabolic stasis that accounts for all of the variability that we observe in cellular response. There are only two sources of change in the cell: synthesis and metabolism. The synthesis pathway is precise and unerring. Thus, all of the systematic and 'random' variability that we see in cellular response, results from 'metabolic stasis' equations similar to equation 7. And ultimately, that variability has only one source: variability in environmental factors. Because this is a reversible thermodynamic system, its next step is always and everywhere governed by the Principle of Least Action (8). Thus, this next step is unique and can be known with incredible precision.

THE NEXT METABOLIC STEP IS...
...EXACT AND PRECISE

- THE STATISTICAL THERMODYNAMIC TREATMENT ACCORDING TO HELMHOLTZ MAKES THE NEXT METABOLIC STEP EXACT
- APPLYING THE PRINCIPLE OF LEAST ACTION THEN MAKES THE NEXT STEP PRECISE

Rule 5: Because the cell is a region where reversible thermodynamic processes govern, the next step in any cellular process, either synthesis of metabolism, is precisely determined by the Principle of Least Action.

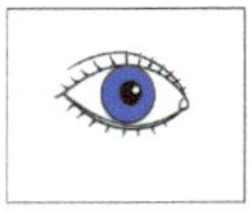

It is now possible to inventory the chemical composition of Reaction Sequences A, B, and C to estimate the total trans-membrane movement of atoms. The results are illustrated in Table 1. The first column is the number of reactions in each reaction sequence. The columns that follow are the number of specific atoms transported across the cell membrane. The final two columns are the number of molecules of oxygen and water transported.

Table 1 – Computed Number of Ions and Molecules Transported Across the Cell Membrane

Reaction	Number of	Number of Atoms or Molecules Transported						
Sequence	Reactions	C	H	O	N	P	H_2O*	O_2*
A	14	28	112	168	-	-	28	28
B	1	2	16	14	2	-	2	3
C	1	2	20	22	2	1	3	3
Hybrid Totals		32	148	204	4	1	33	34

* Tabulated separately. However, these figures are included in the columns labeled O and H.

Eckert and Randall (11) have presented a similar stoichiometric analysis of human blood:

(8)

$$\begin{array}{ccccccccc} C & : & H & : & O & : & N & : & P \\ 32 & : & 212 & : & 86 & : & 4.7 & : & 0.74 \end{array}$$

These authors point out that, "This (stoichiometry) holds for most, if not all, living organisms." This presentation multiplied their stoichiometry by a constant to compare equivalent carbon content. The C:N:P ratios are essentially the same. However, the H:O ratios differ.

I have reasoned that the Table 1 totals are the excess blood constituents involved in respiratory transport, and the difference between the two:

	C	H	O	N	P
(9)	0	64	118	-0.7	0.26
or	0	64	118	0	0

is the combination of H_2O (32 molecules) and O_2 (43 molecules), the later being the amount of oxygen in the blood that is delivered at the tissue level. The Eckert and Randall (11) analysis is for human blood. There could be differences in other cell and tissue types.

From Table 1, the apparent respiratory quotient at a teleost fish gill can be calculated as:

(10) $$Respiratory\ Quotient_{apparent} = \frac{CO_2}{O_2} = \frac{32}{34} = 0.94$$

While the actual respiratory quotient at the tissue level is more correctly found by:

(11) $$Respiratory\ Quotient_{actual} = \frac{CO_2}{O_2} = \frac{32}{43} = 0.74$$

the difference being oxygen that facilitates the overall transport, but ends up as H_2O rather than O_2 delivery to the blood and tissue.

Rule 6: Cells harvest random heat of motion from their environment, converting the harvested energy into ordered radiation domain energy which is stored within covalent bonds.

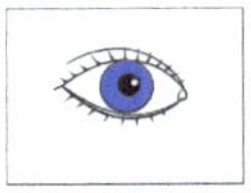

At this juncture, it is important that we clearly understand the physical process that energizes this treatment of respiratory metabolism, and indeed, all metabolic pathways. Its proximate cause is the free energy made available to respiratory metabolism by H_2O_2 molecules. However, at the most fundamental level, it is the action of biological enzymes that systematically 'quiet'

substrate molecules, and in this way harvest random heat of motion from the environment, and converts it to chemical free energy. Thus, Rule 6…

E. The DNA Template, and the Precision of the Synthesis Pathway

We now see how all of biology can be divided into two sub-processes: synthesis and metabolism. Synthesis is the very exact, deterministic process that faithfully follows precisely the same sequence during each cell cycle, or in a broader context, in the synthesis of progeny. It is strictly deterministic, and only experiences variation as a result of mutations, or disease.

Rule 7: Cell synthesis is the very precise sequence of steps that recurr in each cell cycle, resulting in exact copies of proteins, cells, and individuals. Only mutation and disease alter the stepwise synthesis pathway.

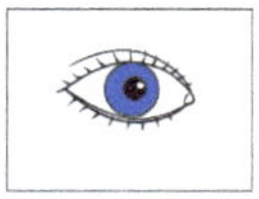

Synthesis takes place on two very different templates. Both are very far-from-equilibrium structures that contain the temporal sequence for the entire mitotic cycle.

The first of these is the DNA template. It contains the precise temporal sequence of all mass associations that will occur during cell synthesis. The transcription process and its temporal development are generalized in the Central Dogma (2) of molecular biology, and are known in extraordinary detail.

The second is an energy template. This is the template of vacancies in the far-from-equilibrium blackbody spectra. It is the skeleton upon which the entire covalent bond structure of the daughter cells will be constructed. It too has a very precise sequence. In particular, we will recall (12) that it is T_R, the radiation temperature that increases monotonically, adding onion-like layers onto the far-from-equilibrium blackbody curve. Each of these layers corresponds to a unit increase in the Wien frequency. This addition of specific quanta is illustrated in the Figure. Thus, at every moment in the cell synthesis cycle, there is one, and only one next covalent bond. It is linked to a single mass association, from among all mass associations that are pending. It is the one covalent bond that satisfies the Principle of Least Action.

These two templates operate simultaneously. We might even conjecture that each serves as a check on the other, insuring the fidelity of the overall synthesis process. The mass synthesis sequence is set by DNA transcription, while the Principle of Least Action specifies the precise energy levels to be associated with that next mass association.

BLACKBODY SPECTRAL ENERGY ACCUMULATION (in cells)

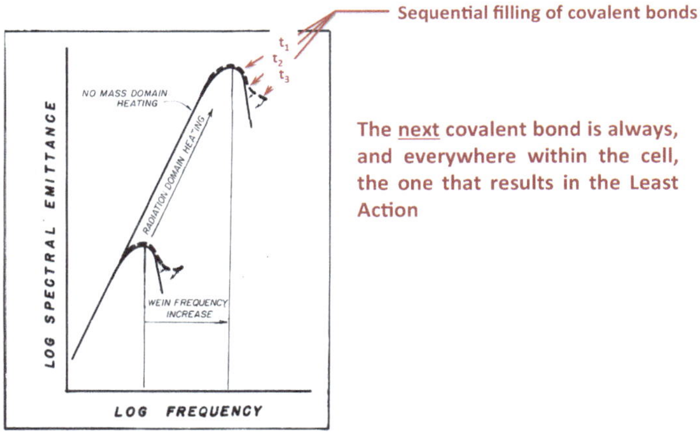

F. Cell Model Essentials

At this juncture, we need to digress, and look at the broader mathematical modeling context that our cell model will have to conform to. This will help us assess the adequacy of our knowledge base for model development, and as it turns out isolate the one feature that we still need before we even begin model development.

A mathematical model always has four components: 1) The system's geometry or structure, 2) Mechanisms for transport and storage of mass and energy, 3) A kinetic system consisting of the model's state variables, and the rate constants governing their transformations, and 4) The forcing functions that make the model progress in time and space.

BIOLOGICAL MODEL ELEMENTS

- SYSTEM GEOMETRY
- TRANSPORT MECHANISMS
- SYSTEM KINETICS
    - STATE VARIABLES
    - RATE CONSTANTS/VARIABLES
- FORCING FUNCTIONS

Let's take an inventory of our model relative to each of these. The geometry of our cellular system is well defined. At its most basic level, it consists of an aqueous milieu outside a cell membrane or cell wall, and aqueous interior consisting of a high dielectric solvent: pure water;

and organelles each set off by their own membranes, and at the interior of each, another aqueous solution. Carrying our description further we find large and small mass associations, or molecules, that have very specific sequences and spatial orientations. DNA's structure is an example. We can even talk about how these molecular associations change over time. These are the cell characteristics that form the geometry of the cell model that we hope to develop.

The model transport describes the spatial relocations that occur within this geometry. We spoke earlier about the transport of simple ions across a cell membrane. That discussion even made estimates of the relative numbers of atoms and molecules that crossed the membrane, and provided a means for estimating the derivatives describing ion and molecular transport. Similarly, transport occurs across organelle membranes, and ultimately within the mitotic spindle, as two daughter cells evolve. But, whereas there was a lot of certainty about the cell's geometry, the transport appears to be less well defined. For instance, is it the random motion of molecules and ions that results in their appearance at the active site of an enzyme, or is there some more sophisticated mechanism moving essential reactants to places within the cell where they are needed. Similarly, is the energy within the cell localized in its molecules, or is there a more global energy structure that allows energy exchange to occur over distance.

Model kinetics have two parts. The first consists of the model's state variables, while the second describes the mechanisms that change these state variables in space and time. The obvious state variables are the inorganic and organic molecular constituents of the cell. These number in the hundreds of thousands, and are partitioned into the bulk aqueous cell milieu, or its organelles. They constitute an enormous bookkeeping task if they are to be modeled individually. However, as we have seen, the modeling task can be simplified to three state variables, if instead, we model the energy structure within the cell and its organelles. This understanding is what is missing from current methods in quantitative biology.

The second part of the model's kinetics consists of the rate constants for changes to the state variables. These rates are modified by the concentrations of other state variables as in the case of Monod kinetics, or directly by other state variables as in the case of temperature dependencies. Biochemical kinetics are fairly well known, at least if we consider biochemical change involving only a few reactants and their product(s). The inventory of specific reactions taking place in different kinds of cells is gradually being catalogued and sequenced. Rates are being associated with many of these reactions, as are the experimental factors that appear to modify the reaction rates. But now we are getting into a grey area. If we ask how these reactions are propelled forward in time, or suspended, we default to talking about signals based in the DNA, RNA, and proteins. In fact we might even introduce into the conversation 'messenger' molecules or other nebulous signal mechanisms. True, the molecules that we attribute these functions to, do exist. But are they really causative, or merely other intermediaries that are being pushed through time with everything else within the cell membrane. The fact that over all these years of research, reliance on these tricks has not led us to any fuller understanding, might be suggesting that we need to move on. I say this because, when all of our cards are on the table, we cannot say anything about what makes this cellular engine go. That brings us the fourth model component, its forcing function(s).

Forcing functions are the factors that add mass or energy to the model, and thereby induce change. In Newton's model, gravity is the predominant forcing function, although others come into play depending upon model circumstances. For example, trajectory models might include thrust, wind speed and direction, air resistance, skin friction, and remaining fuel mass to modify speed and location. A model of enzyme activity employs temperature, substrate concentration, Monod rate modifications, pH and redox state. In the Newtonian tradition, we need to define those forcing functions that cause modifications to the cell, and ultimately impart temporal progression to it. Ideally this is a singular variable that all cell function is dependent upon. Those of us who have been down this road, know that this is not an easy task. Indeed, this is the juncture at which cell modeling invariably breaks down.

G. Time's Arrow in the Evolution of Cellular Processes

At this point, we have developed enough understanding to form a plausible hypothesis regarding the cell's forcing function(s). Our cell model advances on two fronts. The first is the very precise, deterministic evolution of cell synthesis. It is a singular process that reads/writes two templates simultaneously: the DNA template is read, and the far-from-equilibrium blackbody spectral template of vacancies is filled. The second front on which the cell advances is a statistical thermodynamic process that defines cellular metabolism. It is a self-regulating mechanism that always and everywhere results in the Least Action step. This is because the process that we are looking at is thermodynamically reversible. What we are now looking for is the singular process that drives both of these sub-processes in the forward time direction.

We will begin this discussion by amending our definition of the cell. In particular, we will add to our earlier definitions by noting that the living cell is a highly reduced state of matter. I will use pe (15), the negative log of the electron activity (analogous to pH), as a measure of the redox intensity at any location within the cell. pe might be looked at, for the moment, as a measure of electron density, and will exist at different intensity within the various cell organelles. In general, we can say that there is a definite gradient in pe between the aqueous medium outside the cell, and its nuclear, or other organelle, centers. Basic metabolic functions are theorized to take place at relatively low pe, while progressively higher functions require ever higher redox states. Thus, whatever is causing the redox state within the cell to increase is probably a good candidate for the forcing function that propels the cell forward in time.

THE CELL IS...

...A HIGHLY REDUCED STATE OF MATTER

In Section C, we hypothesized a respiratory metabolic process in an aerobic animal cell wherein electrons originating in an organic nutrient source, reduce oxygen to form the high-energy intermediary, H_2O_2. This then mediates what, for illustration purposes, I have characterized as three, coupled respiratory reaction sequences. These reactions are distinctive in the way that they pump electrons across the cell membrane, and in so doing, continually increase the intracellular electron density, and with it, the intra-cellular redox intensity. It is entirely possible that this active transport of electrons into the cell is the singular forcing function that we seek.

Let's step back and look at this possibility from a practical standpoint. It makes sense that the cell's forcing function is embedded in the metabolic pathway. This forcing function is the single most important cell function. Thus, it makes sense that it should occur where the cell has its maximum ability to respond to environmental change. It is also sensible that any requisite part of the forcing function be available on demand. The large oxygen reservoir in aerobic and aquatic environments meets this standard.

If we think about it for a moment, it also makes sense, at least in retrospect, that oxygen uptake is causally linked to the cell's forcing function. Oxygen deprivation results in an abrupt cessation of the cells forward progression in time. This suggests a role in the forcing function mechanism.

However, relative to oxygen uptake by the cell, aerobic animal cells are opportunistic in their nutrition, and thus, their electron needs. When nutrition is available, it is incorporated into the organism where it undergoes a continuous, but slow digestive process. When there is no opportunity to feed, stored food continues to be digested, and in this way, the electron source remains sufficient. Indeed, the cell can survive for long periods without nourishment because of the relatively long time scale over which digestive processes occur; on the order of hours or even days. Here we are looking at natural selection at its most fundamental level: the conditions required to make early cells viable.

These mechanisms facilitate a continuous supply of the cellular forcing function's two essential elements: oxygen and electrons. It is appropriate that this is an active transport system. Passive transport would be too unreliable for such an important cellular process. It is also more satisfying this way. We have seen how atomic constituents are transported into the cell in the precise stoichiometry that its metabolic and synthesis processes require. The energy requirement is augmented by capturing waste heat of motion, thereby making the transport feasible and extraordinarily efficient. Virtually no excess energy is consumed or produced. No laws of physics have been violated. The overall process is operating at the very limit of the Second Law.

I want to make one other observation before leaving this discussion of the cell's forcing function. In particular, it appears to me that there is another reason why oxygen is so important. It is one of a handful of molecules that exhibit large differences in free energy between the formulations of Gibbs and Helmholtz. It is this property alone that makes the coupled metabolic Reaction Sequences A, B, and C possible. These reaction sequences cannot move forward if only the Gibbs free energies are considered. It is only in using the Helmholtz free energies, and in so doing, affirming the assumption that cellular processes are exclusively in the domain of reversible thermodynamics, that this cell model can move forward on its own.

Now let's return to our discussion of redox potential as the forcing function for our cell model, and focus on the deterministic process of cell synthesis. We find that the electron transport process brings electron pairs across the cell membrane, not as individual pairs, but rather the large number of electron pairs demonstrated in equation (7) and resulting from the aggregate transport of the three reaction sequences. This has the effect of incrementally increasing the intra-cellular redox potential, and with it the forward progression of both metabolism and synthesis. From a cell synthesis standpoint, we have filled another vacancy in the far-from-equilibrium blackbody spectra, and facilitated another covalent bond in the synthesis pathway.

However, from the standpoint of our far-from-equilibrium blackbody theory we have done something more. We have increases the cell's redox potential, and according to our definition, we have increased the electron activity that pe measures. This can be seen more clearly if instead of pe, we look at its analog in our far-from-equilibrium blackbody theory, the radiation temperature, $\boldsymbol{T_R}$.

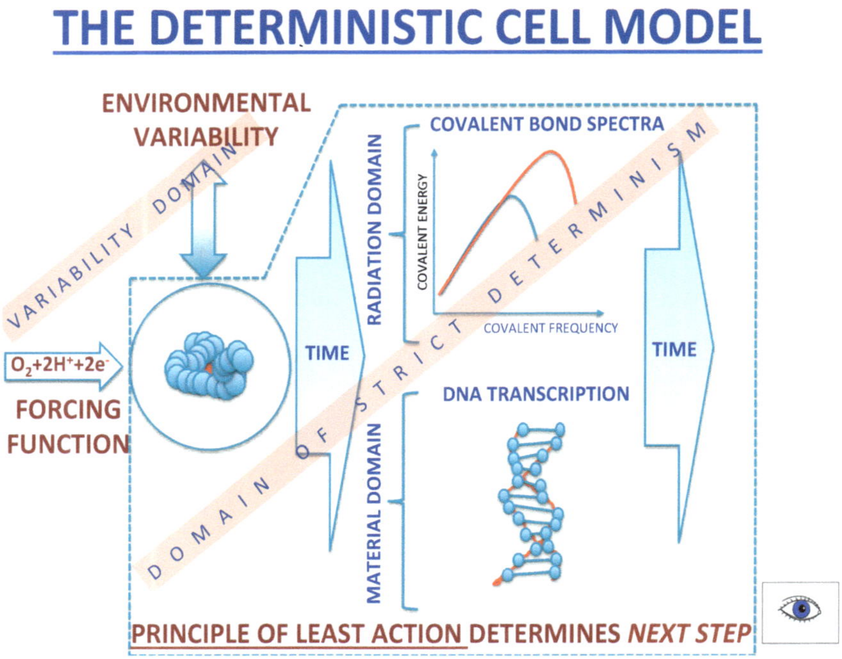

H. Temperature is a Derivative

Temperature is a derivative (8). It is, by definition, the rate of energy emittance across the boundary of any free body. At equilibrium the rate of absorbance and emittance are equal and either rate defines the temperature. Our case is a little different. We will recall that according to the far-from-equilibrium blackbody theory a covalent bond is made up of two electrons sharing a single electro-magnetic quanta, wherein the photon always remains in its reversible thermodynamic state, being emitted by one covalent electron, absorbed by the other, and re-emitted before it can be transformed into its irreversible state; electric charge. You will also recall that we can quantify the temperature of either covalent electron by summing the individual emittance quanta in a time interval, $\boldsymbol{\Delta t}$, arriving at the derivative $\boldsymbol{h\nu^2}$. If we now set $\boldsymbol{\nu}$ equal to the Wien frequency, $\boldsymbol{T_R \cong h\nu_m^2}$, is the measure of electron activity that pe represents. We will talk about Mossbauer Resonance and solar temperatures in Chapter 5.

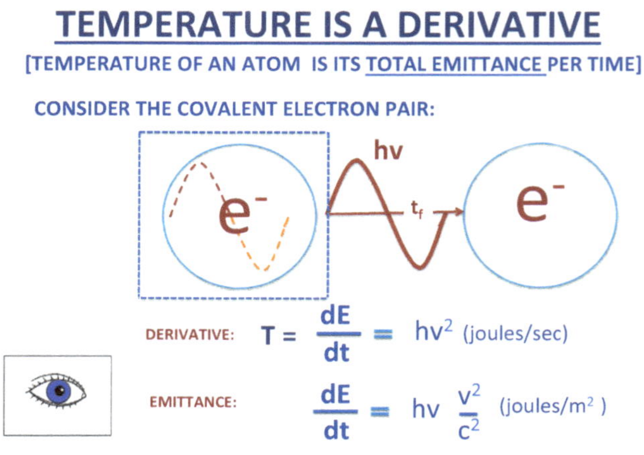

As electrons are brought across the cell membrane their density increases, and sequential increments in the Wien frequency occur. The energy of the covalent bond formed at that instant assumes the next highest energy level in the far-from-equilibrium blackbody spectra and ratchets up the redox state within the cell a small, but quantifiable amount. The new covalent bond is that defined by the Principle of Least Action.

SOLAR TEMPERATURES ARE POSSIBLE

NOW CONSIDER TWO NUCLEI IN MOSSBAUER RESONANCE:

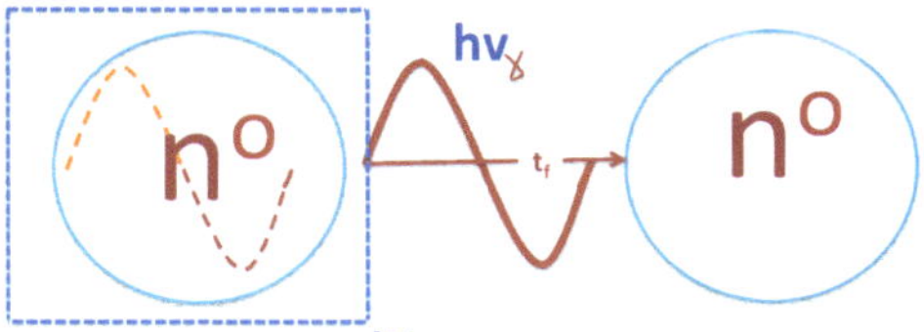

DERIVATIVE: $T = \frac{dE}{dt} = h\nu_\gamma^2$ (joules/sec)

EMITTANCE: $\frac{dE}{dt} = h\nu_\gamma \frac{\nu_\gamma^2}{c^2}$ (joules/m^2)

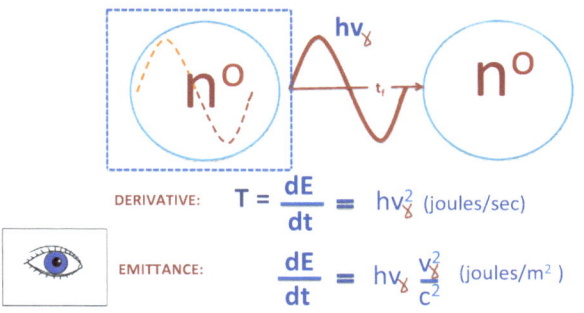

Not only has this resulted in a new covalent bond, but the energy field within the cell, what I have referred to as redox state, increases slightly everywhere within the cytoplasm (or within another cell organelle), causing changes always in accordance with the Principle of Least Action. For example, as the redox field intensity increases, the conformation of amino acid chains changes to bring about an equivalent internal electrical field strength. This of necessity involves folding inward to bring the density of electrical field strength within the molecule into stasis with the ever-increasing ambient redox intensity (i.e. the ambient radiation temperature, or ambient pe). Static charges within the enzyme protein modify the protein condensation in exactly the same way that it happened the last time that this cells predecessors were at this same redox state. Thus, the very exact folding pattern for the amino acid chain is achieved step-by (very small) step. This happens simultaneously with all amino acid chains, with each active enzyme achieving maturity in its proper sequence, and being active within its specific redox range(s).

Rule 9: Enzyme folding into its active conformation follows a specific sequence of redox mediated steps. This is a global process effecting all amino acid chains simultaneously, and bringing them into their active conformation sequentially, and in a precise order. This happens simultaneously among all amino acid sequences within the cell or cell organelle.

This is Monod's *ultima ratio*.

I. Open for Discussion

This manuscript is, at this point, complete in its original intent. All that remains is a final chapter where the same Theory of Heat is applied to another heat problem in physics: a theory of cold fusion, or as I prefer to call it: the Least Action Nuclear Process. This is presented as a verification of the model in an entirely unrelated scientific discipline.

But before we go there, I want to briefly share with you some of my own thoughts on other areas of the physical and biological sciences where I believe this same Theory of Heat might provide useful insights. These are questions that are tangentially related to our discussions thus far, but where I have not had the time to fully develop the theory. You can look at these as suggested assignments. Please send me your essays. Write simply, follow my formats, and let's see if we can finish this book.

I. In inorganic chemical systems, change of phase typically occurs as a function of temperature and pressure. For example, when water transitions from ice to liquid water, it is because the mass domain energy within the ice lattice becomes saturated, and additional heat input partitions into both the radiation domain, and also, the next available phase where molecular motion can be stored; the liquid state. Also think about the transitions to vapor and plasma states, and how the pressure enters into your theory.

II. The division of biological function into its two components: synthesis and metabolism might allow us to give new insight into another problem that I have enjoyed thinking about over the years. Why did dinosaurs become extinct? Consider the metabolic form in this Chapter's Equation 7. A similar, but different equation might have been operative at the level of the dinosaur's respiratory interface with its environment, perhaps including a gas that became limiting, or excessive, over time. Equations of this type can have very sharp pK's: transition point where, suddenly, a small change in one of the independent variables results in abrupt changes in one of the state variables. Think chemical titration. Consider an environment where life has evolved in response to an atmospheric composition that differs in one or two significant ways from today's atmosphere. Now transition that environment to the post-Cretaceous period (today's atmospheric conditions). What might the Cretaceous respiratory metabolic function have looked like, and where might the relevant pK have operated to cause species shifts to new respiratory fundamentals.

III. I have speculated that aging, cell size, and the cancer phenotype are all tied to cell synthesis function. Consider the possibility that we might be able to consider human memory in light of metabolic Equation 7. Perception might then be considered a metabolic response to environmental signals that set up a precise Least Action sequence of holistic chemical patterns suggested by Equation (7). Memory might then be re-cognition of the same Least Action cascade…a place where the present moment's sequence of markers all line up as they did at the remembered moment, thereby triggering the same feeling or cognition. Where can we go with this concept?

IV. Finally, it would be fun to apply this Theory of Heat to sono-luminescence theory, something that was alluded to earlier.

J. References

(1) Szumski, D.S., Theory of Heat II - A Model of Cell Structure and Function, unpublished manuscript, 2003.

(2) Crick, F., Central Dogma of Molecular Biology, Nature 227:561-3, 1970.

(3) Monod, J., Chance and Necessity, Alfred A. Knopf, NY, 1971.

(4) Maxwell, J.C., A Treatise on Electricity and Magnetism, Clarendon Press, Oxford, 1892.

(5) Gibbs, Willard J., On the Equilibrium of Heterogeneous Substances, Connecticut Academy of Science, Transactions 3:5, P1-08-248, 1873.

(6) Newton, I, Mathematical Principles of Natural Philosophy, 1726, Translated by Andrew Motte, 1729.

(7) Maren, T.H., Carbonic Anhydrase; Chemistry, Physiology, and Inhibition, Physiological Reviews 47:595, 1967.

(8) M. Planck, Eight Lectures in Theoretical Physics-1909, translated by A.P. Wills, Columbia U Press, NY, 1915.

(9) Szumski, D.S., et al, Evaluation of EPA un-ionized ammonia Toxicity Criteria, J. Water Pollut. Control Fed, March, 1982.

(10) Szumski, D.S., The Mechanisms of Ammonia Toxicity to Teleost Fish, with Reference to Ion Transport at the Gill Membrane, Unpublished manuscript, 1981.

(11) Eckert, R., Randall, D., Animal Physiology, W.H. Freeman & Co., S.F., 1978

(12) Szumski, D.S., Theory of Heat I - Non-equilibrium Blackbody Radiation Equation, unpublished manuscript, 2000.

(13) Prigogine, I., From Being to Beginning, W.H. Freeman and Company, San Francisco, 1980.

(14) Nicolis G., Prigogine I., Self-Organization in Nonequilibrium Systems, John Wiley and Sons, NY, 1977.

(15) Stumm, W., Morgan, J.J., Aquatic Chemistry, John Wiley & Sons Pub., New York, NY, 1970.

Chapter 5

Heat Model Verification - Modeling The Cold Fusion Process

A. Introduction

During the last two decades it has become evident that low energy nuclear reactions are occurring in Fleischmann-Pons (F-P) electrolytic cells (1). These reactions are unprecedented in nuclear physics, and are at least for now, hidden from understanding because a suitable theoretical framework has not been forthcoming. (2)

'New Science' is Required

It is theorized that excess heat (3,4) and ^4He (5) are generated, and that the heat evolved is consistent with the mass difference (5,6) in the reaction:

(1) $(2)_1^2H^+ \Rightarrow {}_2^4He + 23.9\,MeV$ (ignition requirement= 0.01MeV)

It is also becoming apparent that the reactions taking place are a near surface phenomenon (7) that is spatially clustered (7), occurs in bursts (7), and has a cyclic character within those bursts (8). Even more controversial than the contention that Reaction (1) occurs, is a growing body of data showing other nuclear transmutations in F-P cells (9,10,11,12,13), and still more alarming, in living cells (14).

The degree to which new physics underlies these experimental observations is not known. But, among theoreticians it is considered more likely that the present conundrum will be resolved by extensions of known physical principles, perhaps in ways that we cannot immediately imagine.

This research endeavors to provide insight into three theoretical issues. First, recognizing that the fusion reaction's energy has its origin within the experimental apparatus, we explore a mechanism for accumulating the energy required by Reaction (1) or other fusion/fission reactions. Second, there has to be a way of storing that energy within the apparatus, and in some way, disguising it until the moment of ignition. And the third is the elusive coherence principle that focuses the accumulated energy on specific nuclear transformations, and not others. The goal here is to show how a different view of heat processes, one that includes both irreversible and reversible thermodynamics, might inspire a comprehensive cold fusion theory.

B. Theory of Heat

Heat exists in two domains that continually exchange energy as any arbitrary thermal system tends toward new quasi-equilibrium states. These are the domain of molecular motion, and the domain of heat radiation. The first might be referred to as the mass domain. Its description was first formalized by Maxwell (15), and then by Boltzmann (16). Their theory represents the molecular velocity distribution of an ideal gas as a function of the system's temperature and the gas molecules' mass. It is an equilibrium theory stating the functional dependence of temperature and thermal motion. It was Helmholtz who had first shown that molecular motion is equivalent to heat; an observation that is central to what follows. Max Planck, in his 1909 lectures at Columbia University (17), elevates this insight to an equal footing with Maxwell's treatment of light as electromagnetic waves.

Heat energy also exists in the radiation domain. The theoretical framework describing equilibrium conditions there bears the revered names of Rayleigh, Wien, and Planck. Planck's equation (18) describes the equilibrium temperature dependence of blackbody spectral emittance.

Reversible thermodynamic processes are believed to be rare in nature. These are processes that produce a net zero free energy change, and are described by the thermodynamic treatment of Helmholtz, but not that of Gibbs. In all cases, reversible processes can be completely described by the Principle of Least Action. A discussion of this principle and the thermodynamics of reversible processes are presented by Planck (17).

In a previous paper (19), I proposed a mathematical form for the blackbody spectral distribution that permits a glimpse into its non-equilibrium, and far-from-equilibrium characteristics. The theory treats light absorption as a two step process: the first (and this will be the important one to what follows) being wholly reversible, the second irreversible and entropic. In special cases only the first step occurs, and the energy absorption is adiabatic, that is, without loss of Joule heat. The functional form of the non-equilibrium blackbody spectra is given by:

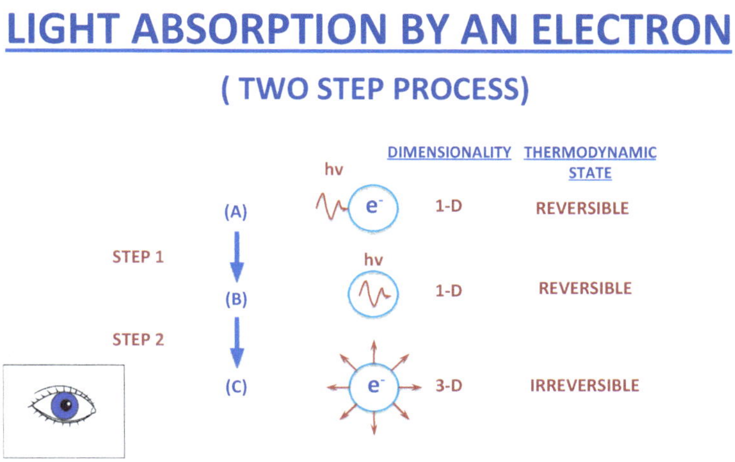

(2) $$K(\upsilon_1) = \frac{e_\upsilon}{a_\upsilon} = k_b T_m \cdot \frac{\upsilon_1^2}{c^2} \cdot \frac{1}{e^{f_3(\upsilon_1/\upsilon_m)}}$$

Blackbody Spectra = = Emittance (Rayleigh Law) / Absorptance (This study)

where

(3) $$f_3\left(\upsilon_1/\upsilon_m\right) = \left(\upsilon_1/\upsilon_m\right)^2 \left[\frac{1}{2}\left(\ln\left(\frac{\upsilon_1}{\upsilon_m}\right)\right)^2 - \frac{3}{2}\ln\left(\frac{\upsilon_1}{\upsilon_m}\right) + \frac{7}{4}\right]$$

(4) $$\nu_m = 5.89\times10^{10} \cdot T_R \ ({}^oK)$$ (Wien frequency),

and $K(\nu_1)$ exists where the number of quanta is equal to or greater than 1.

Planck's equation for the equilibrium case is given by:

(5) $$K'(\nu_1) = h\nu_1 \frac{\nu_1^2}{c^2} \cdot \frac{1}{e^{\frac{h\nu_1}{k_b T}} - 1}$$

The non-equilibrium form of the equation is close to, but not exactly Planck's at equilibrium.

Steady State Blackbody Spectra

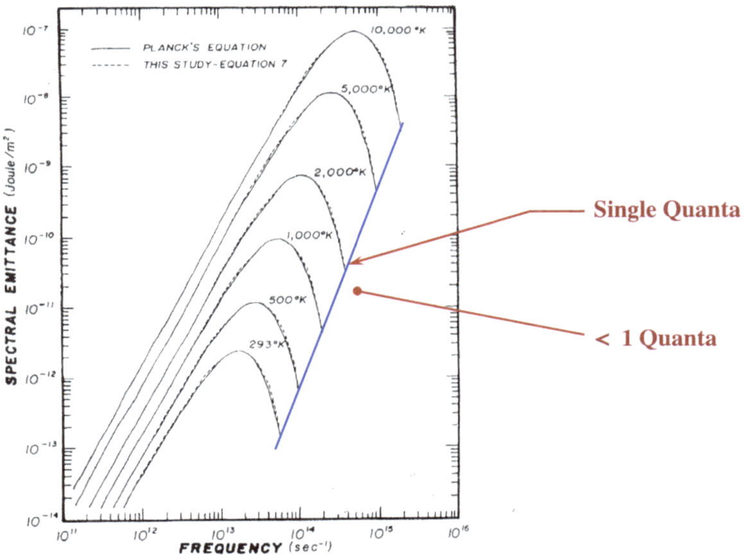

Secondly, the theory suggests that independent temperature scales might represent the mass and radiation domains. The figure illustrates the principle characteristics of the non-equilibrium, or more accurately the far-from-equilibrium, blackbody radiation spectra. Two equilibrium cases are shown: 300^0K and $100{,}000^oK$. Curve A, labeled Mass Domain Heating, refers to the transient initial condition where heating is initiated by increasing molecular motion, for example, by frictional input of heat. The Wein Frequency remains constant momentarily, and there is a logarithmic increase in the spectral energy at all frequencies.

FAR-FROM-EQUILIBRIUM BLACKBODY SPECTRA

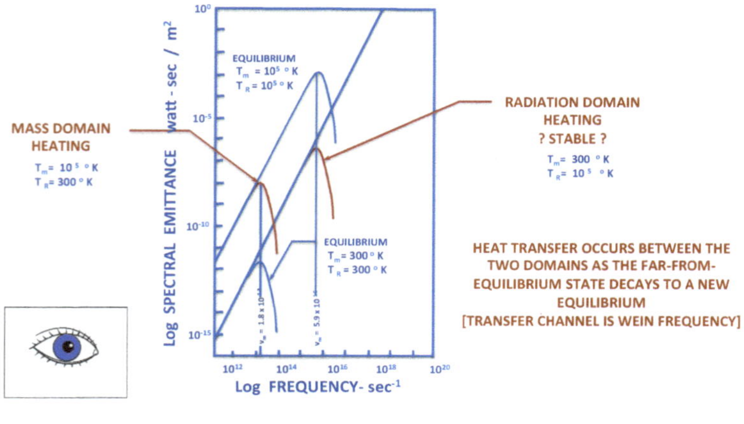

If an identical amount of heat is instantaneously added via the radiation domain alone (by introducing higher energy radiation), the thermodynamic temperature, T_m , is initially constant, and the Wien frequency increase, shifts the emittance spectra to higher frequencies as shown in curve B.

Both of these cases decay to an equilibrium spectrum similar to, but with higher total energy than, the initial equilibrium case. At the new equilibrium condition, the mass and radiation temperatures become identical, and it is not possible to determine from which of the two domains the original heating took place. However, as will be shown in what follows, there are circumstances under which the second of these spectra might be held in its far-from-equilibrium condition, and in this way store vast amounts of energy in a nickel or palladium cathode that is apparently at about $60\,^{o}C$.

This paper attempts to reconcile this non-equilibrium theory of heat with experimental observations from cold fusion experiments, and show how the energy stored at room temperature can be made accessible to fusion reactions.

C. Consider A Thermodynamically Reversible Process

In thermodynamically reversible chemical processes all of the available energy, including that of thermal motion, is utilized, and none of the energy in the post-reaction space is lost to random thermal motion. I have found that the easiest way to understand the principles involved in what is to follow, is by considering the reaction occurring at the active site of a biological enzyme. The reactants and enzyme have fundamentally different thermodynamic properties. The former are relatively small molecular forms having both chemical potential and thermal motion. The enzyme, on the other hand, is a large molecule, that in its globular active form, has very little or no thermal motion, and an undetermined chemical potential.

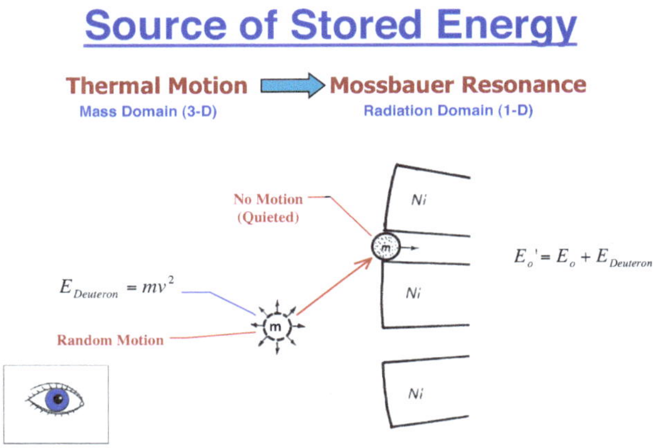

The enzyme mediated biological reaction brings reactant molecules into conformational position, generally by electro-static attraction, and in so doing, makes improbable reactions, probable. The secret lies in transformations to the energy states of both the reactants and the enzyme. In particular, cleavage at the active site eliminates thermal motion in the reactants; it quiets them. The First Law tells us that the 'lost' thermal energy must be conserved, and in its limit, the Second Law tells us under what conditions the reaction can proceed. In essence, the reactants' thermal motion has become part of the reactant-enzyme complex, elevating its overall free energy content.

If the thermodynamics of the enzyme/reactant complex are truly reversible, the total energy is passed on to the reaction products, and there is no energy residual that contributes to thermal motion in the reaction space. As long as these conditions are met, the reaction proceeds in accordance with the Principle of Least Action. Then conformational changes occur, and the product becomes subject to the slightly altered thermal state of its environment. In essence, the enzyme has harvested random heat motion from the environment, converted it to useful work, and in so doing increased the radiation domain's heat content. What at first appears to be a violation of the Second Law, is simply its limiting case, a zero net energy reaction that produces a more negentropic state.

It was Szilard's argument (20) concerning Maxwell's sorting demon (15) that correctly showed how the negentropy stored by the demon as molecular organization and intellect, sponsors his trick, in apparent violation of the Entropy Principle. In the case considered here, it is the massive information content in the globular enzyme form that allows the demon to operate. In a related theoretical context, Prigogine (21) would label the enzyme/reactant complex a dissipative structure: a far-from-equilibrium thermodynamic state, which once formed, allows no recourse to the previous state, and in so doing, lowers the local entropy.

D. Consequences of Deuterium Absorption into a Metallic Lattice

Consider a deuterium ion, (${}^{2}_{1}H^{+}$), in Fleischmann and Ponn's original experiment (1). Its total energy is the sum of its chemical potential and its kinetic energy, ε, that associated with temperature dependent random motion. The kinetic energy is given by the product of the deuteron's mass, m_d, and its temperature dependent velocity squared:

(6) $$\varepsilon = \frac{m_d(v(T_m))^2}{2} =$$

Where: $m_{d=} m_p + m_n = 3.34\times10^{-27}\ kgm$, and v is the velocity of an individual deuteron, or in our simplified treatment, the average velocity of an ensemble of deuterons at F-P cell temperature, T_m. We will assume an average velocity of 0.2m/sec, a simplification that ignores for the moment the system's actual velocity distribution, but facilitates illustrative calculations.

The average kinetic energy of the deuterons in their F-P cell can be calculated as

(7) $$\bar{\varepsilon} = 6.68\times10^{-29}\,Joules = 6.68\times10^{-22}\,ergs.$$

When a deuteron first encounters the nickel matrix, it is absorbed into it in a process that we will assume to be thermodynamically reversible, and similar to the enzyme process described above. The deuteron is 'quieted' to zero velocity, and zero kinetic energy. The First Law requires that the kinetic energy be conserved in the metal hydride lattice. And because the loading process is thermodynamically reversible, the energy storage is adiabatic, with no losses to Joule heat.

To place an order of magnitude estimate on this energy storage, we will use the 0.2cm diameter x 10cm electrode from Fleischmann and Pons 1989 experiments (1). The surface area of the cathode is 6.28×10^{-4} meters. Assuming β-phase absorption approximating $\beta - PdD_{0.85}$, and having a lattice parameter of 0.405nm, the number of filled sites at the surface of the cathode, ξ, is approximated as:

(8) $$\xi = {6.28\times10^{-4}m^2}/{(0.405\times10^{-9}m)^2} = 3.83\times10^{15} \text{ surface sites}$$

The 85% load factor yields: 3.25×10^{15} $[^2_1H^+]$ sites on a single atomic layer at the cathode surface. We assume that the total cathode is immersed in heavy water.

The energy storage capacity, E, of only the surface layer of atoms in this cathode is:

(9) $$E_{surface} = \bar{\varepsilon}\cdot\xi = 2.2\times10^{-6}ergs = 1.35MeV,$$

which is more than sufficient to ignite the fusion reaction:

(10) $$(2)^2_1H^+ \Rightarrow {}^4_2He + \gamma_{23.9MeV}$$ (ignition requirement = 0.01MeV)

Absorption of deuterons into the second, third, and deeper atomic layers in the cathode, increases the total energy availability proportionally. For example, ignition is achieved in a single cathode layer if the average deuteron velocity is reduced to 2.0cm/sec, and in 100 layers if the velocity is further reduced to 0.2cm/sec. Energy accumulation increases at higher temperatures, and doubles again if ${}^{2}_{1}H_2$ molecules form within the interstitial space (22). Thus, deuterium's sequestered thermal motion appears to be more than sufficient for ignition.

E. Possible Modes of Energy Storage within a Metallic Ni Lattice

The mechanism presented thus far has the advantage of providing qualitative insights into several theoretical issues. First, it provides a simple explanation of how the ignition energy is first acquired in the Ni cathode. All that might be required is a reversible thermodynamic process that harvests kinetic energy from the F-P cell environment. Secondly, it provides a basis for understanding the 'breathing' mechanism in the SRI experiments (8). Stored thermal energy and deuterium are expended and need to be replaced on a periodic basis. This could manifest as a harmonic superimposed on the excess heat output. Third, it offers an explanation of the apparent surface nature of the effect. This is where the energy accumulation occurs, and where it is most probably utilized and renewed. And, finally, it provides a plausible explanation of why loading rates increase at higher current density/temperature. The total energy storage per mole of ${}^{2}_{1}H^{+}$ is increased as the square of the average deuteron velocity.

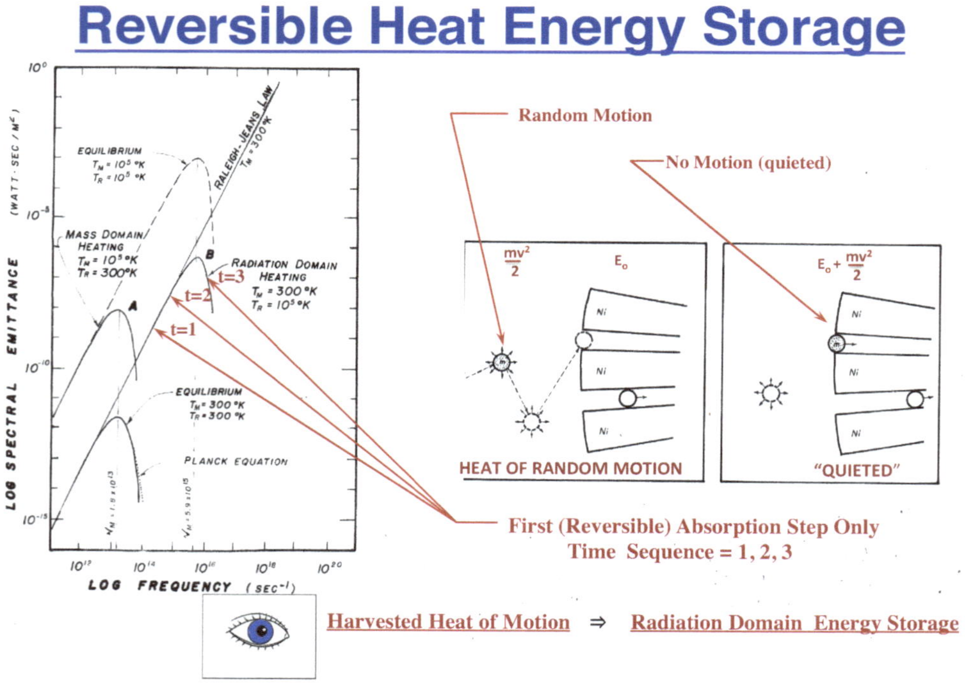

For this transfer to be a thermodynamically reversible one, the lattice energy has to increase sequentially, in discrete amounts, exactly equal to each sequestered deuteron's total kinetic energy. Then it must be held there in opposition to all entropic tendencies until ignition.

How is this energy stored during the loading phase of the experiment? We will begin by assuming that deuterium loading is a singular, multi-site, reversible process. The energy is not dissipated in incremental amounts, as it was in the enzyme example. Instead, the Ni lattice has massive numbers of active sites that must be filled before the reversible process can proceed further. During this loading, no energy is lost to thermal motion. Thus, the energy stored is either entirely in the radiation domain, or it moves from the mass and radiation domains of heat energy to another energy type where it can be held in a completely reversible state.

This constraint, that the energy storage be in a thermodynamically reversible state, allows us to further limit the possibilities. The mode of energy storage could be 1) electro-magnetic, in which case the energy of, for example, discrete metallic bonds might be increased by quantum amounts forming covalent bonds or excited electronic states; or 2) it could be magnetic energy storage in paramagnetic Pd's electron spin re-orientation, or 3) it could be energy stored as excited nuclear states. It is not stored as elastic stress, electric charge, or atomic vibration, all of which are entropic processes. In addition, the energy storage mechanism must make allowances for energy storage that spans a continuous range from the ambient temperature of the experimental apparatus, through thermonuclear temperatures.

It appears to me that the best explanation for the lower bound might be found in energy storage within discrete covalent bonds; each covalent electron pair alternately absorbing and emitting electro-magnetic energy that remains in a wholly reversible state, i.e. the first step of the two step, photon absorption process. This is a stable far-from-equilibrium state (upon which excited electron states might be superimposed). It

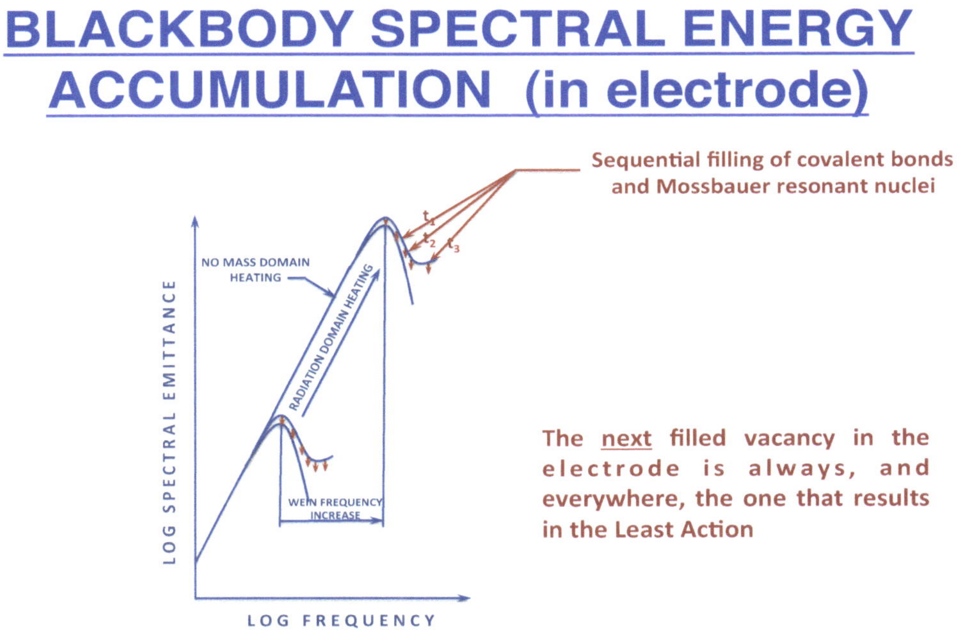

manifests as an increase in the cathode's redox potential. As the total energy storage increases further excited nuclear states become active, ultimately bringing the reversibly stored cathode energy to gamma levels, where Mossbauer resonance, a reversible process, prevails, and energy storage occurs as resonant gamma exchange.

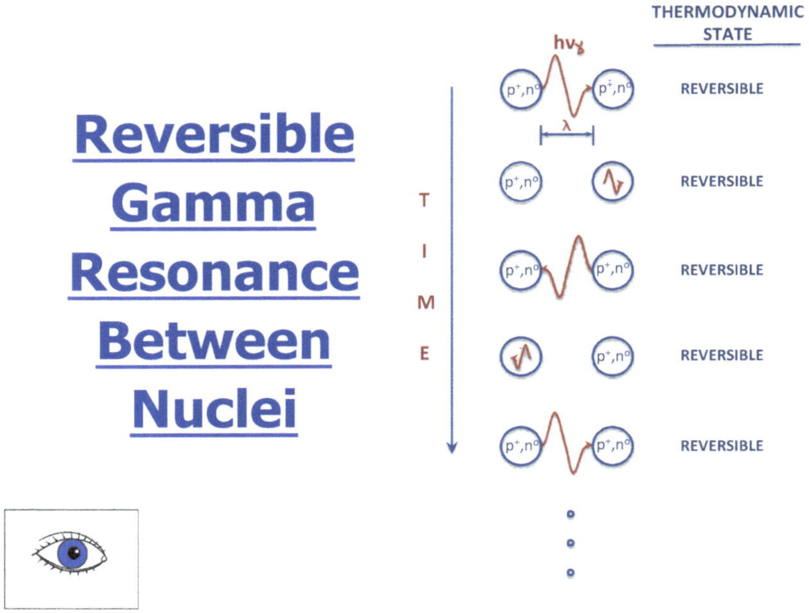

If we now look more closely at the consequences of energy storage in excited nuclear states, we find that this energy is stored entirely within the atomic structure of the lattice, and without any external manifestation. No heat energy is emitted. The thermodynamic temperature remains unaffected by the deuterium loading, and in this way, the process's energy storage is masked from observation. The observer witnesses a very typical electrolysis apparatus, and has no hint of the continually increasing radiation temperature within the lattice's atomic structure.

The spectra labeled B in the Reversible Heat Storage figure represents the distribution of energy levels corresponding to this storage of heat energy. These are filled sequentially at each Wien frequency. Then the Wien frequency increases one unit, and another layer is added to the spectral structure. Eventually, the Wien Frequency reaches gamma intensities, and the radiation temperature approximates that in the solar core, about $10^7\,{}^oK$ as illustrated in the figure. The figure contrasts the temperature regime (T_m and T_R) that this theory postulates, to that in the solar core. It suggests that the energy spectra required for ignition in the Tokamak includes both the indicated radiation domain energy and also the mass domain kinetic energy, and is about four orders of magnitude higher than that operative in the F&P cell. The total energy requirement is many orders of magnitude greater. In essence, the cold fusion process takes an energy shortcut around the enormous kinetic energy required for thermonuclear fusion. In this way, we see that the cold fusion process is actually quite hot.

Far-From-Equilibrium Radiation Domain Energy Storage

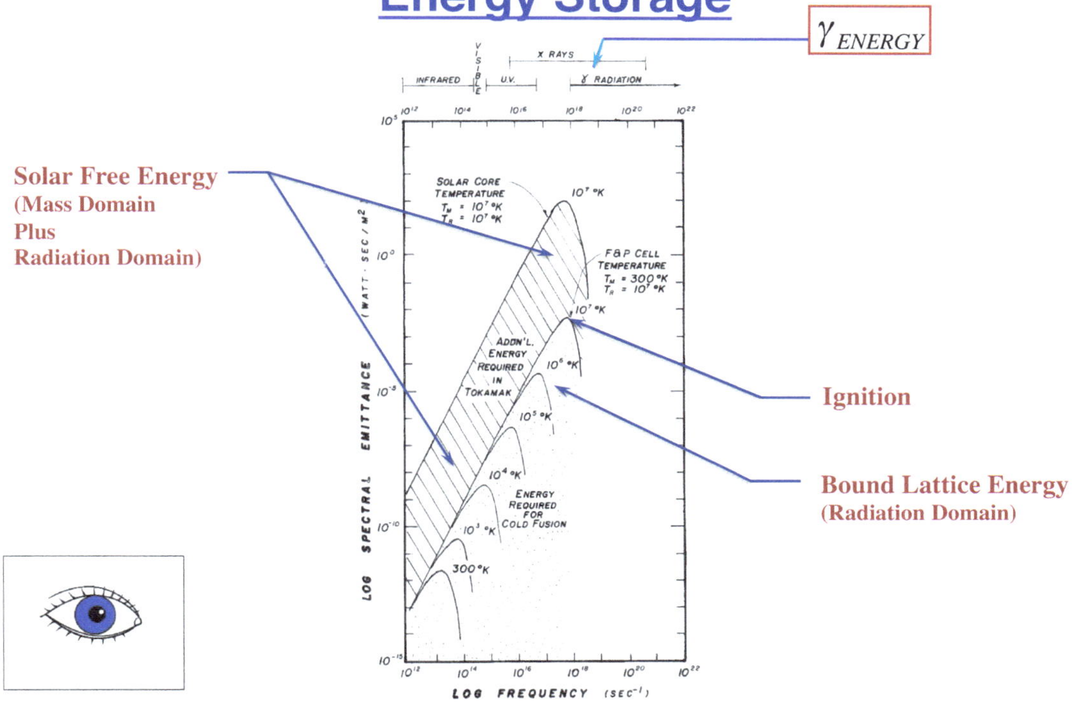

Let us, for the moment, assume that fusion and/or fission events occur in our electrode as T_R approaches $10^{7\,o}K$. Where are the Gamma emissions?

We can answer this question by recalling that we are dealing here with an extension of blackbody theory wherein electromagnetic energy of all wavelengths is emitted and fully absorbed within the lattice. The mass quantities involved in this absorption and emission are electrons at the low energy end of the spectrum, and atomic nuclei as the energies increase through gamma intensities.

Where are the Gamma Emissions?

- **Blackbody spectra - All wave lengths are absorbed and emitted**
  - Low Energy - Covalent Bonds
  - High Energy - Mossbauer Resonance
- **Mass Changes occur as Emission/Absorption in Blackbody Spectra**
- **Gamma Emission and Absorption are:**
  - Intrinsic to this System
  - Internal to It
  - In Effect, Masked by It

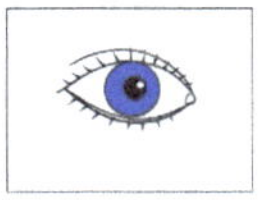

This is an absorption/emission process wherein electro-magnetic energy is shared between identical mass quantities. In effect, gamma emission occurs as part of the normal blackbody dynamic, and within that context, participates in nuclear fusion or fission events. Gamma absorption and emission adds and subtracts energy quanta to/from the spectrum. Because this is a metal lattice, the emission/absorption occurs in accordance with Mossbauer kinetics, without recoil or heat loss, or more precisely in a completely thermodynamically reversible manner. And, as long as there is room in the spectra, the gamma energy released by fusion and fission events is fully absorbed elsewhere in the lattice by a nucleus having exactly the same ground/excited state as the emitting nucleus.

This is simple blackbody behavior, occurring now, at a very far-from-equilibrium state. One might rightfully ask: is it the mere fact that the gamma portion of the blackbody spectra is being filled, that causes the nuclear reactions? Or, is it the radiation temperature that sponsors fusion/fission events?

One consequence of associating the energy requirement for the nuclear reactions with the blackbody spectra is that all reactions that can occur, do. This is because the entire continuum of spectral energies is available in the far-from-equilibrium blackbody spectra. At the end of the next section, we will see how the 'next', nuclear reaction is selected from among all possible reactions that are pending at any one moment in time.

F. Experimental

Now let's return to the mechanisms of thermonuclear fusion/fission under these conditions. Miley's data from electrolysis of nickel-coated micro-spheres (9) provides a suitable data set for analysis. I have inventoried what I believed are the most likely nuclear reactions occurring in the nickel coated micro-spheres, which I will refer to as the electrode. Those reactions that consistently produced observed transmutation products without producing extraneous isotopes are presented in Appendix B. I have used 'isotope of [element]' data extracted from Wikipedia (23) to complete the energy calculations in these tables.

Candidate nickel-deuterium and electrode impurity-deuterium nuclear reactions are tabulated in the first column. The second and third columns are the initial isotopes formed, and the final stable product of its decay. I initially thought that the reversible portion of the nuclear reaction would extend only to column 2, and that the heat evolved from the experimental apparatus would be that from beta-decay of the initial fusion product to stable isotopes. I also suspected that the isotopes observed in the 'post-experiment' electrode would be all of the decay products of the initial fusion/fission reaction. This worked fairly well as long as I made some other assumptions.

First, I had to allow gaseous products to 'gas out' of the apparatus. This seemed reasonable, but then I had to look at the half lives of gaseous intermediaries, to make the judgment call – is it reasonable to assume that this gaseous product had enough time to 'gas out'? Wherever unstable gaseous products were formed in the initial fusion/fission reaction, the half-life of the isotope is provided in the table so that the reader can assess the opportunity for gassing that product out of the electrode before it reacts further. Stable gaseous products were assumed to gas out of the electrode.

Another source of concern at this point was the feasibility of fusion reactions involving 5, 6, ...10 deuteron. Having completed hundreds of decay sequences for all types of possible nuclear reactions, it had become apparent that multiple deuteron reactions were the only way to produce most of the low atomic weight products in Miley's Table. But there were practical problems here. Did the multiple deuteron reactions occur all at one time? Or, do they occur sequentially? There seemed to be a proximity issue in a face centered cubic lattice if the reaction involved, for example, 10 deuterons. Yet this still seemed a preferred route, because sequential deuteron addition produced many short half-life, radioactive isotopes that probably were not available for further deuteron addition. Another concern about multiple deuteron reactions is apparent in the reaction involving ${}^{58}_{28}Ni$ plus six deuterons, yielding ${}^{70}_{34}Se \xrightarrow{\beta^-} {}^{70}_{33}As \xrightarrow{\beta^-} {}^{70}_{32}Ge$. But, ${}^{70}_{32}Ge$ is not measured in the post-experiment electrode. It is 'absent'. Is this because the addition of six deuterons to ${}^{58}_{28}Ni$ never occurs, or occurs only after the first five additions (, which could be all that occur during this experiment's duration), and is not occurring in sufficient amounts to be measured yet in the experiment. Or perhaps, ${}^{70}_{32}Ge$ undergoes fission to $(2){}^{39}_{17}Cl \uparrow$. This result is ambiguous. There are many reactions involving 6 or more deuterons that yield stable terminal isotopes that Miley did measure. I also saw that although the final fission product, chlorine gas, is a plausible fate for the unwanted ${}^{70}_{32}Ge$, why doesn't every other final, stable isotope undergo fission.

There are also questions regarding the initial amounts of specific isotopes available for reaction. For example the nickel isotopes in the initial electrode are probably present in the normal isotopic composition ${}^{58}_{28}Ni$ (68%), ${}^{60}_{28}Ni$ (26%), ${}^{61}_{28}Ni$ (1.1%), ${}^{62}_{28}Ni$ (3.6%), and ${}^{64}_{28}Ni$ (0.9%), indicating that reactions involving ${}^{58}_{28}Ni$ are far more likely to produce measurable quantities of fission/fusion products as those involving ${}^{61}_{28}Ni$ or ${}^{64}_{28}Ni$. Electrode impurities should react according to this same concentration dependance.

Another issue that crops up in the data analysis presented here is illustrated in the fusion of ${}^{64}_{28}Ni$ with 3, 5, and 7 deuterons. In each case, there are multiple reaction pathways. Is one path preferred over the other? Why is one of the product isotopes absent (${}^{70}_{32}Ge$, ${}^{78}_{34}Se$) even though it occurs along an overwhelmingly preferred pathway?

I have also looked at the range of fusion reactions between the initial electrode isotopes (i.e. ${}^{58}_{28}Ni + {}^{60}_{28}Ni$, ${}^{58}_{28}Ni + {}^{107}_{47}Ag$, or ${}^{107}_{47}Ag + {}^{68}_{30}Zn$), and also the full range of those fusion reactions, but incorporating one or more deuterons (i.e. ${}^{58}_{28}Ni + {}^{107}_{47}Ag + n({}^{2}_{1}H^{+})$). These pathways produce large numbers of stable isotope products that Miley did not observe, as well as some that were observed.

These are the kind of questions that have kept me up at night.

Overall the tables show that the reactions producing the lower atomic weight portion of the final electrode composition are: 1) fusion reactions of initial electrode isotopes with one or more deuterons, 2) fission reactions of initial electrode isotopes or 'absent' isotopes, or 3) alpha decays. I have also looked at the same three types of reactions, but involving the products of the initial reactions. This was less productive.

The first three columns in the table summarize my initial analysis of Miley's data. It shows that my methods to this point account for all of the Miley isotopes through ${}^{75}As$, about half of those in the atomic mass range of 76 through 125, and none of the higher mass isotopes ${}^{126}Te$ through ${}^{208}Pb$. I had originally suspected that I would find pairs of reactions that produced a net zero mass change. That is, that there would be no net change in energy content in the initial reaction of the reaction sequences when two or more coupled reactions occurred simultaneously. This would satisfy the reversibility constraint. However, I now realize that net zero energy changes are not required for the reactions to satisfy the reversibility requirement. All that is required is that the blackbody spectra has an absorption and emittance quantum of exactly the same energy as the difference between the initial nuclear reaction in the overall reaction sequence, and the final stable products of that reaction. The blackbody form accomplishes this implicitly. The entire continuum of spectral energies is present by definition. Thus, all reactions that can occur, do occur.

The absence of atomic masses above ^{208}Pb and the mass accumulation in ^{206}Pb, ^{207}Pb and ^{208}Pb is informative in several respects. These three isotopes are the radioactive decay end products of the uranium, actinium, and thorium series respectively. These are the most likely, if not the only, paths to them. Thus, it is reasonable to conclude that nuclei having mass greater than Pb-208 are produced in the electrode. There is no other plausible explanation of how such neutron heavy, lead products could form in the electrode given the composition of the initial nickel electrode and its impurities. More important is my estimation that there simply aren't enough neutrons in the initial system. Neutron formation appears to be one of the fundamental processes taking place in the cold fusion electrode.

The only way that heavy, trans-lead isotopes can form is by rapid neutron capture in the kind of nucleosynthesis that occurs in supernovae. I am out of my element here. But, if the radiation temperature hypothesis presented in this paper can be given any weight, it is not a great leap to the conclusion that the radiation temperature, T_R, of our far-from-equilibrium blackbody spectra could also approach stellar supernovae temperatures. Remember, temperature is a derivative. It increases as the frequency of the exchanged quanta, and also because the exchange occurs more frequently as the frequency increases.

Following this argument further, it is particularly noteworthy that no intermediate, radioactive isotopes of the uranium, actinium, and thorium series are present. Decay along these routes produce unstable intermediates that should be detected in cold fusion experiments. Is it possible that the theorized reversible reaction process short circuits the decay steps to a stable end product, producing only a mass/energy change for the overall reaction? In this way none of the radioactive intermediates or time delays associated with long half-lives occur, and cold fusion proceeds without the messy radioactive signatures of other nuclear processes.

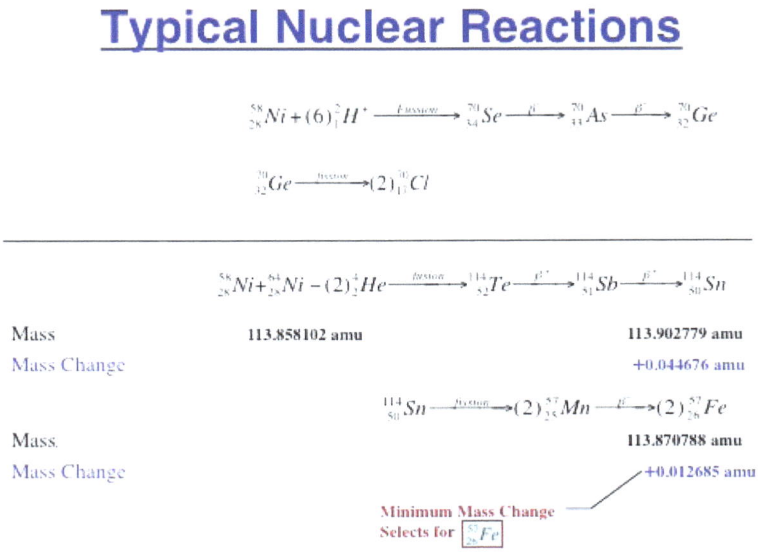

At this point, I was out of ideas and assumptions for explaining reaction products that are absent in Miley's data. Still there were outliers. And the one thing that I known with absolute certainty, is that a law of physics cannot have outliers.

As it turns out, the solution to this dilemma lay in this study's initial premise: the reactions involved in cold fusion processes are thermodynamically reversible. The one thing that all thermodynamically reversible reactions have in common is that their evolution is completely described by the Principle of Least Action. Thus, the one rule governing the selection of specific nuclear isotope products, and the exclusion of others can be summarized.

Rule 1 - All fusion and fission reactions that can occur are candidates. The one that actually produces a product along any reaction pathway is the reaction sequence that satisfies the Principle of Least Action.

The result of applying this rule to the reactions in Appendix B is shown in column 4. I have calculated the mass of the reactants in column 1, subtracted from it the mass in the final stable product(s) (column 3), and shown that difference in column 4. Bold type is used to highlight the product selected for by the least action principle. In several cases, I have also shown other possible reaction paths to illustrate that the selected one does indeed have the least energy change. In all cases except one, where absent products formed, or where there were several stable isotope choices, the Least Action Principle selects for an isotope in Miley's Table 3. This is true regardless of the sign associated with the overall energy change. The table presents the analysis results in order of increasing reaction energy regardless of the sign of the energy change. I call this model the Least Action Nuclear Process (LANP) Model.

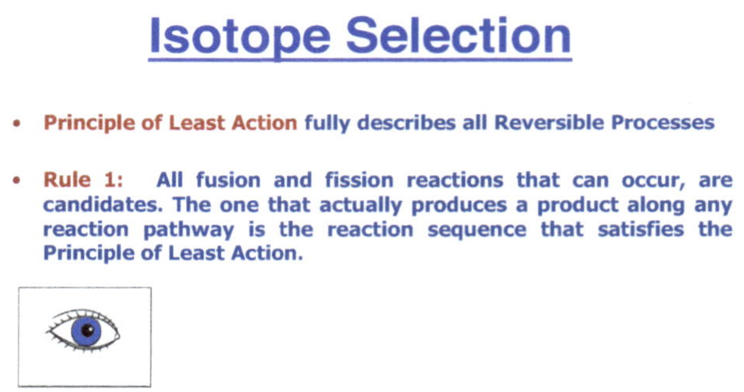
Isotope Selection
Principle of Least Action fully describes all Reversible Processes
Rule 1: All fusion and fission reactions that can occur, are candidates. The one that actually produces a product along any reaction pathway is the reaction sequence that satisfies the Principle of Least Action.

I have tested this model with an independent set of data that was more challenging than the data set used in its development. In particular, I had already drawn about 20 more complex decay diagrams for fusion reactions involving silver and nickel, and also multiple nickel reactants. These produced initial isotopes in the 120-205amu range. In many cases there were more than 10 intermediate decay products, some with extremely long half-lives, and others having low probability decay paths that would normally produce a small fractional of a percent of the total

Complex Nuclear Reaction

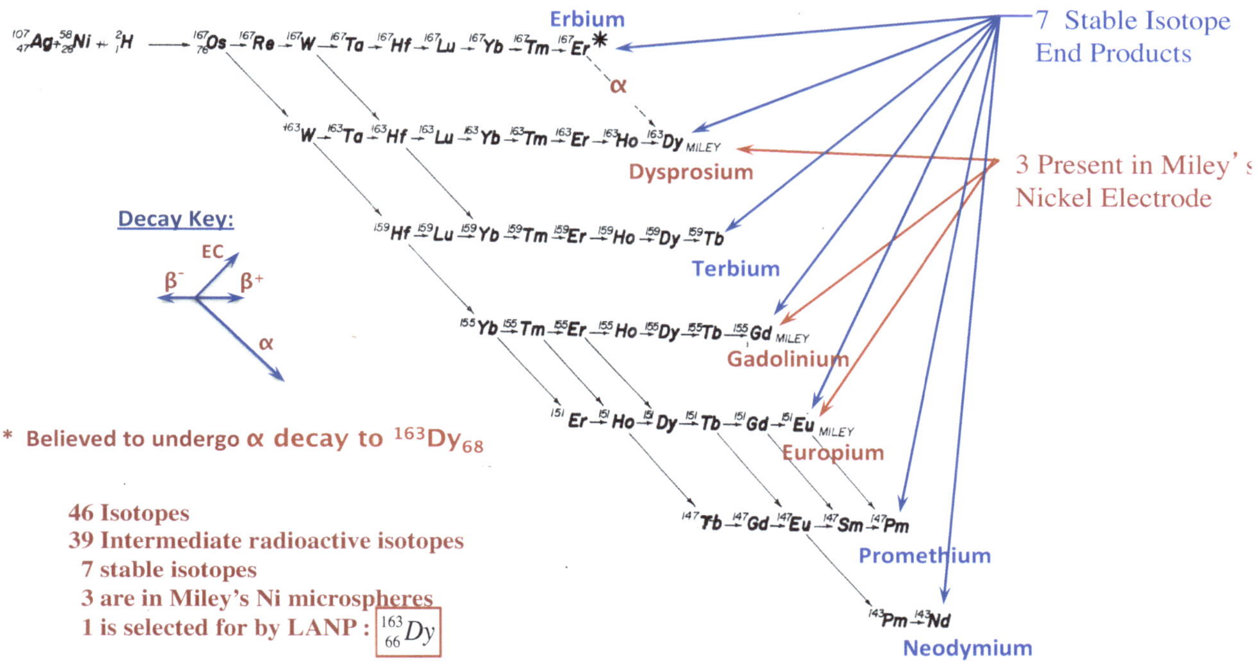

decay product. In all of these cases, the LANP model selects for isotopes in Miley's table, without false positives.

Consider the reaction illustrated in the Figure where the nuclear reaction:

$$^{58}_{28}Ni + {}^{107}_{47}Ag + {}^{2}_{1}H^{+} \xrightarrow{fusion} \ldots {}^{167}_{69}Er \quad \overset{\alpha}{\rightarrow} \quad {}^{163}_{66}Dy + {}^{4}_{2}He$$

mass change → +0.0775065amu +0.0767937amu

produces 45 intermediate radioactive isotopes and 9 stable isotope products, three of which are in Miley's Table 3: $^{151}_{63}Eu$, $^{155}_{64}Gd$, and $^{163}_{66}Dy$. The results obtained from this reaction sequence show how the Principle of Least Action correctly selects for $^{163}_{66}Dy$, but not along the normal decay pathway shown in the Figure. Instead the Principle of Least Action selects for $^{167}_{68}Er$ with a mass change of +0.0775065amu. This is an end product of the normal decay path. It is followed by alpha decay to $^{163}_{66}Dy$, still within the domain of reversible thermodynamics. The energy change drops accordingly to +0.0767937amu, the Least Action product. Helium is produced in this final step, but without generating excess heat via the expected pathway

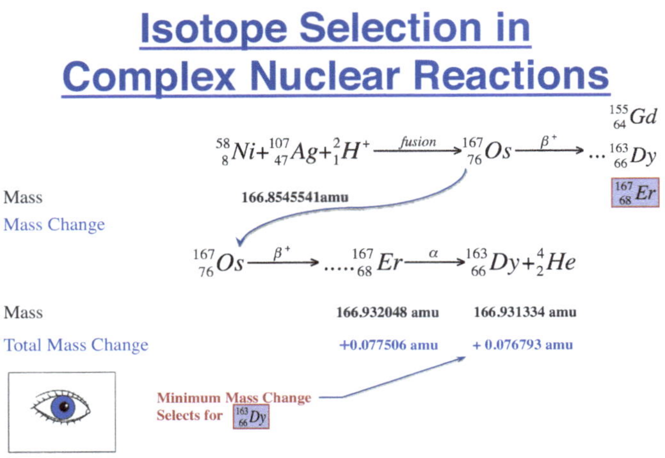

(Equation 1). I call this mode of nuclear decay where no radioactive intermediates are formed, and where there are no half-life time delays, σ-decay.

We are finally ready to look at the issue of excess heat generated in Miley's experiment. The reversibility constraint requires that the overall process be adiabatic. Therefore, we need to explore the limits of that process to identify the step at which it departs from the limiting case of the Second Law where no entropy is produced, and crosses into the domain of irreversibility. First, we note that the mass change appears as a gamma photon in Mossbauer resonance within the far-from-equilibrium blackbody spectra. Negative mass changes increase the total energy within the spectra. Positive changes do the opposite. It seems possible that the summation of these mass changes over all reactions occurring from the point of ignition to the end of the experiment, in some way determines the time history of total excess energy production, or possibly even energy consumption.

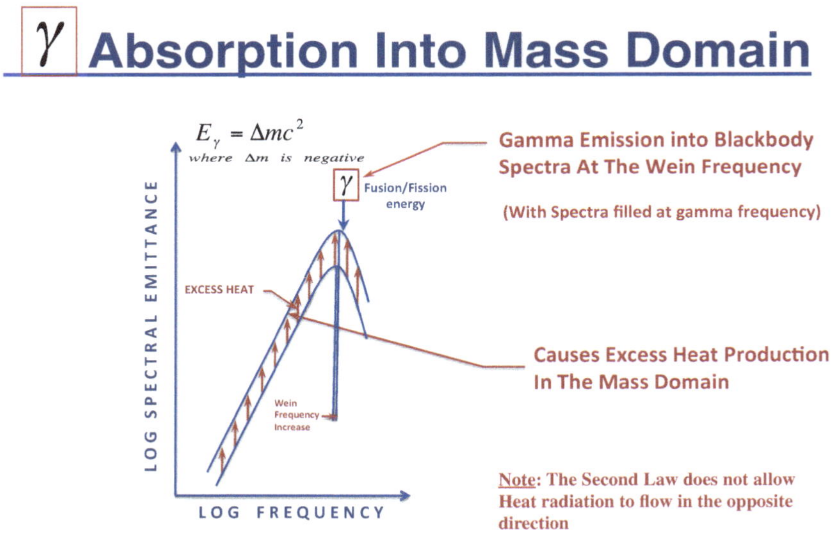

If this energy were to enter the radiation domain at its gamma frequency, but where the number of quanta at that frequency is already saturated, the resulting imbalance necessitates re-distribution into the mass domain. Such an exchange might occur through the Wien frequency channel, where Equation 3 is the transfer function describing the re-distribution of Wien frequency energy into the domain of molecular motion. This increases the thermodynamic temperature, T_m, of the experimental device, producing excess heat. Similar thought experiments describe the evolution of endothermic LANP events, and situations in which the blackbody spectrum does, or does not, have vacancies at the gamma frequency.

LANP - Assertions

- Mechanism for loading energy into the metal hydride lattice
- Mechanism for storing that energy until ignition
- Theoretical basis for the fusion temperature requirement and how it is masked
- Mechanism for selecting products that do and do not occur
- Explanation for the absence of radioactivity

I believe that such a model will ultimately be capable of not only predicting the amount of excess heat produced, but also predicting the sequencing of nuclear reactions and their time history of endo- and exo-thermic contributions to the heat reservoir. I say this because there are only a finite number of possible nuclear reactions for a given initial isotope mix in the electrode, and if the reaction progression proceeds first with an increment in the Wien frequency, and then energy storage in the radiation domain, or energy re-distribution into the mass domain, it should be possible to very precisely map this process. At each next step, its continued evolution is limited to precisely one nuclear event that satisfies the Least Action Principle. This is a very ordered and exact process.

What then is the end point where excess heat production terminates? To place a perspective on this issue, it is necessary to return to the sorting demon discussion in section C. LANP energy is achieved by harvesting random heat of ${}^{2}_{1}H^{+}$ molecular motion, accumulating it in excited nuclear (Mossbauer resonance) states, and then transforming it into thermo-nuclear work. The demon assumes the form of a Ni lattice structure with an affinity for deuterons. By capturing deuterons, and their kinetic energy, he traps heat energy within the lattice, and

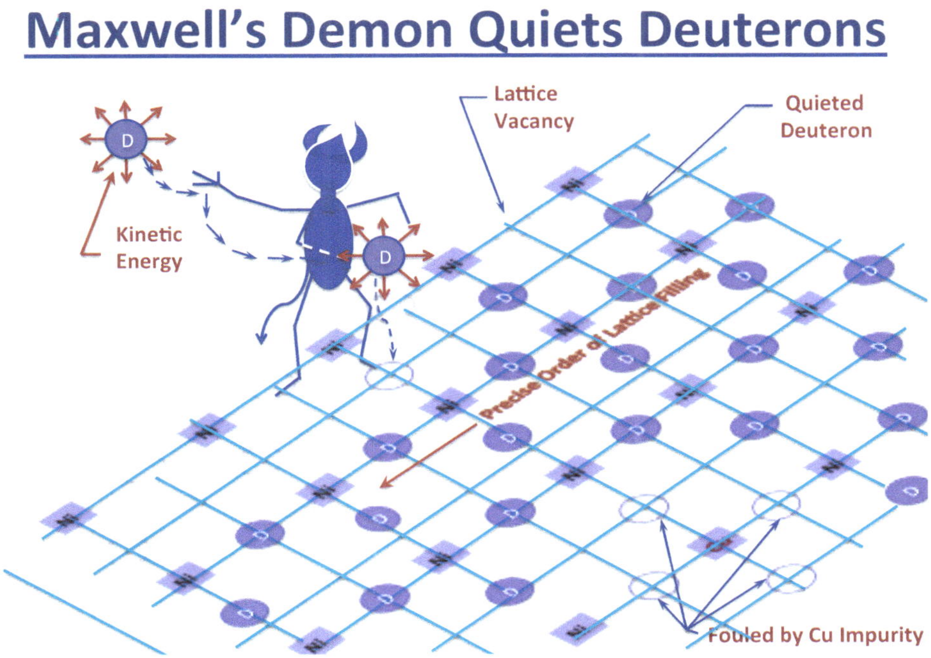

transforms that random heat of motion into excited electronic and nuclear states, and in this way, decreases the electrode's entropy. But, here is a dilemma. What trick has the demon played on us to first cause deuterons to cascade into the palladium lattice? It is this organizing principle that is his real presence, and the secret of this type of neg-entropic process.

Returning now to the cessation of excess heat production, I can only speculate that isotope transformations distort the ability of the near surface lattice to absorb deuterons. In other words, as the near-surface lattice becomes fouled with atoms that no longer participate in the demons trick, the absorption rate decreases. Nevertheless, deuterons continue to occupy deeper sites, albeit at a slower rate, and nuclear reactions continue until the rate of kinetic energy capture becomes rate limiting.

G. Some Final Thoughts

The LANP theory is not nickel specific. It can probably be applied equally well to other metal

LAMP Model Flow Chart

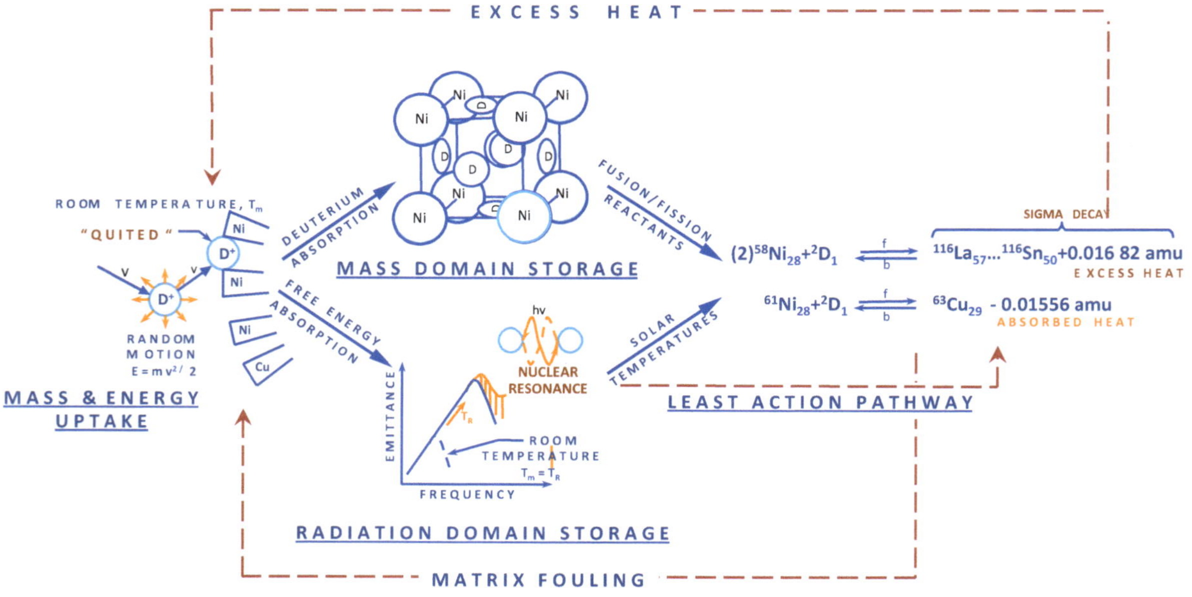

hydrides, and either hydrogen or deuterium absorption. Reactions that can be achieved within the context of reversible thermodynamics occur. Those that do not meet this standard, do not. In fact, there is no reason to limit consideration to metal hydrides. Other processes that are thermodynamically reversible could be subject to a similar theoretical treatment. I would suggest that a good case can also be made that the processes within a living cell might fit into this theoretical framework equally well (24).

The 1-D to 3-D transform function given by Equation 3 appears to be a mathematical statement of the Second Law at the boundary between electrodynamics and mechanics. In its temporal form (19) the equation represents the relative dominance of the forward (entropic) and backward (negentropic) reaction directions. As an example of this function's utility consider its application to the phenomenon of sono-luminescence wherein mechanical energy is converted to electro-magnetic energy. In this case, mechanical energy increases T_m instantaneously without a corresponding increase in T_R . Lacking any mechanism to maintain this far-from-equilibrium condition, the system spontaneously moves toward equilibrium by channeling the stored mechanical energy through the Wein frequency channel, and thence, into the radiation domain. If the energy flux's frequency is high enough, visible light is observed.

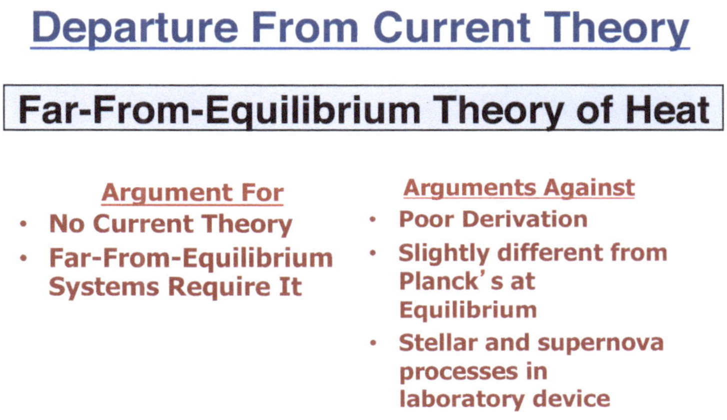
Departure From Current Theory
Far-From-Equilibrium Theory of Heat
Argument For
• No Current Theory
• Far-From-Equilibrium Systems Require It
Arguments Against
• Poor Derivation
• Slightly different from Planck's at Equilibrium
• Stellar and supernova processes in laboratory device

H. Discussion

LANP theory is unique in its ability to describe many of the unexplained phenomena occurring in a Fleischmann-Pons electrolytic cell. These include:

A mechanism for loading energy into the metal hydride lattice.

A mechanism for storing that energy until ignition,

A theoretical basis for the fusion temperature requirement and how it is masked,

A mechanism for selecting reactions and products that do and do not occur,

An explanation for the absence of radioactivity.

The theory also has appeal in that it is not nickel specific, or even metal lattice specific, and it provides a plausible mechanism for the solar temperatures that thermonuclear fusion is known to require. LANP is a very hot process.

The mechanism that causes excess heat to occur requires a detailed methodology for sequencing endo- and exo-thermic reactions, and more discussion of Equation (2), including a more rigorous derivation. Nevertheless, the model demands additional study and experimental work. It answers too many questions to be dismissed.

On the other hand, theoreticians and experimentalists in the field should contain their exuberance for this, or any other promising model. This field is simply too controversial to allow missteps, or premature dialog with the non-scientific community. Places where LANP departs from current theory, and more importantly, from common sense, need immediate study.

Departure From Current Theory

Sigma - Decay

Argument For

- **Observed Isotopes**
- **Absence of Radiation**
- **No Half-Life Delays**
- **Model Calibration Success**

Arguments Against

- **Contrary to Existing Theory**
- **Common Sense**

For example, is it even plausible that all of the intermediate radioactive decay steps, and half-life constrains can be bypassed by LANP's sigma decay process. The absence of any radiation signature in F&P cells, and the observed transmutation products make that conclusion tantalizing. And yet, it is contrary to everything that we currently know about nuclear processes. The same is true of more fundamental aspect of the theory such as its claim of reversibility. This one feature of the theory is without precedent in modern science, and will be attacked vigorously in peer review. Perpetual motion machines, quite simply, are not supposed to exist. Even more implausible is the claim that stellar and supernova processes might occur within a laboratory device.

These claims are almost untenable, and yet they seem to constitute a cohesive theoretical framework that is consistent with the data. We should be very careful not to give LANP too much credibility at this point in its short life, and instead design a scientific plan to achieve rigorous experimental proof one way or the other.

Departure From Current Theory

Reversibility

Argument For

- Principle of Least Action Completely Describes Reversible Processes
- Model Calibration Success

Arguments Against

- Unprecedented in Science
- Perpetual Motion

I. References

(1) Fleischmann, M., S. Pons, M. Hawkins, Electrochemically Induced Nuclear Fusion of Deuterium, J Electroanal. Chem., 261, p. 301 and errata in Vol. 263, 1989.

(2) Hagelstein, et al, Input to Theory from Experiment in the Fleischmann-Pons Effect. In ICCF-14 International Conference on Condensed Matter Nuclear Science, Washington, DC., 2008.

(3) Storms, E., Measurements of Excess Heat from a Pons-Fleischmann-type Electrolytic Cell using Palladium Sheet. Fusion Technol.:23 p 230, 1993.

(4) Fleischmann, M., et al., Calorimetry of the Palladium-Deuterium-Heavy Water System. J. Electroanal. Chem., 287: p 293, 1990.

(5) Miles, M., et al., Correlation of Excess Power and Helium Production during D_2O and H_2O electrolysis using Palladium Cathodes, J. Electroanal. Chem., 346: p.99, 1993.

(6) Miles, M. Correlation of Excess Enthalpy and Helium-4 Production: A Review in Tenth International Conference on Cold Fusion, 2003. Cambridge, MA: LENR-CENR.org.

(7) Mosier-Boss, P.A. and S. Szpak, The Pd/(n)H System: Transport Processes and Development of Thermal Instabilities. Nuovo Cimento Soc. Ital. Fis. A, 112: p. 577, 1999.

(8) McKubre, M.C.H. The Need for Triggering in Cold Fusion Reactions. In Tenth International Conference on Cold Fusion. 2003. Cambridge, MA: LENR_CENR.org.

(9) Miley G., J Patterson, "Nuclear Transmutations in thin-Film Nickel Coatings Undergoing Electrolysis", J. New Energy, vol. 1, no. 3, pp. 5-38, 1996.

(10) Karabut, A.B., Kucherov, Y.R. and Sawatimova, I.B., "Frontiers of Cold Fusion" [Proc. 3rd International Conference on Cold Fusion, Oct. 21-25, 1992, Nagoya, Japan], Universal Academy Press, Tokyo, p.165, 1993.

(11) Bockris, J. O'M, Z. Mineviski, Two Zones of Impurities Observed After Prolonged Electrolysis of Deuterium on Palladium, Infinite Energy Magazine, (#5 & #6), p 67, November 1995.

(12) Dufour, J., Murat, D., J. Foos, Experimental observation of Nuclear Reactions in Palladium and Uranium – Possible Explanation by Hydrex Mode, Fusion Technol., 40: p91, 2001.

(13) Mizuno, T., T. Ohmori, and M. Enyo, Isotropic Changes of the Reaction Products Induced by Cathodic Electrolysis in Pd, J. New Energy, 1996. 1(3): p. 31.

(14) Vysotskii, V., Kornilova, A.A., Samoylenko, I.I., Zykov, G.A., Experimental Observations and Study of Controlled Transmutation of Intermediate Mass Isotopes in Growing Biological Cultures, Journal of New Energy, Vol 5, No. 1, pp. 123-128, 2000.

(15) Maxwell, J, C., Theory of Heat, reprinted Dover, New York, 1871.

(16) Boltzmann, L., Lectures in Gas Theory, Translated by Stephen G Brush, University of California Press, Berkeley, 1964

(17) Planck, M., Eight Lectures in Theoretical Physics, 1909, translated by A.P. Wills, Columbia U Press, NY 1915.

(18) Planck, M., Verhandlunger der Deutschen Physikalischen Gesellschaft, 2, 237, (1900), or in English translation: Planck's Original Papers in Quantum Physics, Volume 1 of Classic Papers in Physics, H. Kangro ed., Wiley, New York, 1972.

(19) Szumski, D.S., Theory of Heat I - Non-equilibrium Blackbody Radiation Equation, unpublished manuscript, 2000.

`(20) Szilard, L., On the Decrease of Entropy in a Thermodynamic System by the Intervention of Intelligent Beings, translated by Anatol Rapoport and Cechilde Knoller, in B.T. Feld and G.W. Szilard(ed), The Collected Works of Leo Szilard- Scientific Papers, MIT Press, 1972.

(21) Nicolis, G, I. Prigogine, Self-Organization in non-equilibrium Systems, John Wiley and Sons, NY, 1977.

(22) Zhang, Z.L., et al, Loading Ratios (H/Pd or D/Pd) Monitored by the Electrode Potential, Abstracts-ICCF-10, Cambridge, MA, 2003.

(23) Isotopes of Hydrogen In Wikipedia, Retrieved 1/04-7/12, from http://en.wikipedia.org.

(24) Szumski, D.S., Theory of Heat II - A Model of Cell Structure and Function, unpublished manuscript, 2003.

Appendix A

Derivation of :

$$f_3\left({}^{v_1}\!/_{v_m}\right) = \left({}^{v_1}\!/_{v_m}\right)^2\left[\frac{1}{2}\left(\ln\left(\frac{v_1}{v_m}\right)\right)^2 - \frac{3}{2}\ \ln\left(\frac{v_1}{v_m}\right) + \frac{7}{4}\right]$$

$$f_n\left({}^{v_1}\!/_{v_m}\right) = \iint \left[\ln\left({}^{v_1}\!/_{v_m}\right)\right]^2 d\left({}^{v_1}\!/_{v_m}\right)$$

Substituting:

$\left({}^{v_1}\!/_{v_m}\right) = x$

first integration:

$$\int (\ln x)^2\, dx = x(\ln x)^2 - 2x(\ln x) + 2x$$

$$\therefore f_2 = \quad x(\ln x)^2 - 2x(\ln x) + 2x \qquad 2^{nd} \text{ Integral}$$

second integration:

$$\iint (\ln x)^2 = \int \left[x(\ln x)^2 - 2x(\ln x) + 2x\right] dx$$

$$\int x(\ln x)^2 dx = \frac{x^2}{2}(\ln x)^2 \; - \int x(\ln x)\, dx$$

$$-\frac{1}{2}x^2(\ln x) \quad +\frac{1}{4}x^2$$

$$-\int 2x(\ln x)dx = \quad -\frac{2}{2}x^2(\ln x) \quad +\frac{2}{4}x^2$$

$$+\int 2x\, dx = \quad +\frac{2}{2}x^2$$

$$\therefore f_3 = \quad \frac{x^2}{2}(\ln x)^2 \quad -\frac{3}{2}x^2(\ln x) \quad +\frac{7}{4}x^2 \qquad 3^{rd} \text{ Integral}$$

Appendix B

Analysis of Miley's LANP Data for Nickel Microspheres

[Presented in order of increasing Least Action]

Table 1 - Nuclear Reactions from Miley's Nickel Data – Ordered in LANP Order

Nuclear Reaction	Initial Isotope	Stable Isotope	Energy Change (amu)
$(2)^{60}_{28}Ni \xrightarrow{fusion}$	$^{120}_{56}Ba$	$^{116}_{50}Sn, ^{118}_{50}Sn, [^{120}_{51}Te]$ **absent**	+0.0424482
$^{120}_{51}Te \xrightarrow{fission}$		$(2)^{60}_{28}Ni$	**0.0000000**
$^{62}_{28}Ni \xrightarrow{fission}$	$(2)^{31}_{14}Si$	$(2)^{31}_{15}P$ **absent**	+0.0191782
$^{62}_{28}Ni - ^{4}_{2}He \xrightarrow{\alpha}$		$^{58}_{26}Fe$ **absent**	+0.0075337
$^{62}_{28}Ni + ^{2}_{1}H^{+} \xrightarrow{fusion}$		$^{64}_{28}Ni$ **absent**	-0.01448087
$^{62}_{28}Ni + (2)^{2}_{1}H^{+} \xrightarrow{fusion}$		$^{66}_{30}Zn$ **absent**	-0.3051524
$^{62}_{28}Ni$		$^{62}_{28}Ni$ **(note 6)**	**0.0000000**
$(2)^{61}_{28}Ni \xrightarrow{fusion}$	$^{122}_{56}Ba$	$^{118}_{50}Sn, \{^{122}_{52}Te\}$ **absent**	-0.0409319
$^{122}_{52}Te \xrightarrow{fission}$		$(2)^{61}_{28}Ni$	**0.0000000**
$(2)^{60}_{28}Ni + (3)^{2}_{1}H^{+} \xrightarrow{fusion}$	$^{126}_{59}Pr$	$^{125}_{52}Te, [^{126}_{54}Xe \uparrow]$	**0.0003958**
$(2)^{58}_{28}Ni + (3)^{2}_{1}H^{+} \xrightarrow{fusion}$	$^{122}_{59}Pr$	$^{118}_{50}Sn, ^{120}_{52}Te, ^{121}_{51}Sb, [^{122}_{52}Te]$ **absent**	-0.0099461
$^{122}_{52}Te \xrightarrow{fission}$		$(2)^{61}_{28}Ni$	-0.050878
$^{122}_{52}Te - ^{4}_{2}He \xrightarrow{\alpha}$		$^{118}_{50}Sn$	**+0.001162**
$^{62}_{28}Ni + (1)^{2}_{1}H \xrightarrow{fusion}$	$^{64}_{29}Cu$	$^{64}_{28}Ni$	-0.014480
		$^{64}_{30}Zn$ **absent(note 1)**	-0.013304
$^{64}_{30}Zn \xrightarrow{\beta^{+}\beta^{+}}$		$^{64}_{28}Ni$	-0.014480
$^{64}_{30}Zn \xrightarrow{fission}$	$(2)^{32}_{15}P$		+0.005368
$(2)^{32}_{15}P \xrightarrow{\beta^{-}}$		$(2)^{32}_{16}S$	**+0.001696**
$(2)^{32}_{16}S \xrightarrow{fission}$		$(4)^{16}_{8}O$ **Example**	+0.037212

$(2)^{58}_{28}Ni \xrightarrow{fusion}$ $^{116}_{50}Sn \xrightarrow{fission}$	$^{116}_{56}Ba$	$^{111}_{48}Cd, ^{112}_{50}Sn, ^{114}_{50}Sn, ^{115}_{50}Sn, [^{116}_{50}Sn]$ **absent** $(2)^{58}_{26}Fe$	-0.0310552 **-0.0020673**
$^{109}_{47}Ag - (1)^4_2He \uparrow \xrightarrow{\alpha}$ $^{105}_{46}Pd \xrightarrow{fission} {}^{52}_{23}V$ $\xrightarrow{fission} {}^{53}_{23}V$	$^{105}_{45}Rh$	$^{105}_{46}Pd$ **(note 2)** $^{52}_{24}Cr$ $^{53}_{24}Cr$	**+0.002936** -1.021133 **H** +0.097915 **H**
$^{107}_{47}Ag - (1)^4_2He \uparrow \xrightarrow{\alpha}$ $^{103}_{45}Rh \xrightarrow{fission} {}^{51}_{23}V$ $\xrightarrow{fission} {}^{52}_{24}Cr$ $^{103}_{45}Rh \xrightarrow{fission} {}^{50}_{23}V$ $\xrightarrow{fission} {}^{53}_{24}Cr$	$^{103}_{45}Rh$ **(stable)**	$^{103}_{45}Rh$ **absent(note2)** $^{51}_{23}V$ $^{52}_{24}Cr$ $^{50}_{23}V$ $^{53}_{24}Cr$	**+0.003039** } -0.018026 } -0.017053
$^{64}_{28}Ni - (1)^4_2He \xrightarrow{\alpha}$	$^{50}_{26}Fe$	$^{60}_{26}Ni$	**+0.0054234**
$^{107}_{47}Ag - (4)^4_2He \uparrow \xrightarrow{\alpha}$ $^{91}_{40}Zr - {}^4_2He \xrightarrow{\alpha}$	$^{91}_{39}Y$	$^{91}_{40}Zr$ **absent** $^{87}_{38}Sr$	+0.010961 **+0.005834**
$^{109}_{47}Ag - (2)^4_2He \uparrow \xrightarrow{\alpha}$ $^{101}_{44}Ru \xrightarrow{fission} {}^{50}_{22}Ti$ $\xrightarrow{fission} {}^{51}_{22}Ti$	$^{101}_{43}Tc$	$^{101}_{44}Ru$ **(note 2)** $^{50}_{22}Ti$ $^{51}_{23}V$	**+0.006036** -1.009963 **H** +0.968373 **H**
$^{59}_{28}Ni - (1)^4_2He \xrightarrow{\alpha}$	$^{55}_{26}Fe$	$^{55}_{25}Mn$	**+0.0063016**
$^{60}_{28}Ni - (1)^4_2He \xrightarrow{\alpha}$	$^{56}_{26}Fe$	$^{56}_{26}Fe$	**+0.0067543**
$^{58}_{28}Ni - (1)^4_2He \xrightarrow{\alpha}$	$^{54}_{26}Fe$	$^{54}_{26}Fe$	**+0.006870**
$^{61}_{28}Ni - {}^4_2He \xrightarrow{\alpha}$	$^{57}_{26}Fe$		**+0.0069412**
$^{61}_{28}Ni - (1)^4_2He \xrightarrow{\alpha}$	$^{57}_{26}Fe$	$^{57}_{26}Fe$	**+0.0069412**
$^{107}_{47}Ag - (2)^4_2He \uparrow \xrightarrow{\alpha}$ $^{99}_{44}Ru \xrightarrow{fission} {}^{49}_{22}Ti$ $\xrightarrow{fission} {}^{50}_{22}Ti$	$^{99}_{43}Tc$	$^{99}_{44}Ru$ **absent(note 2)** $^{49}_{22}Ti$ $^{50}_{22}Ti$	+0.006366 } **-0.007229**

${}^{99}_{44}Ru \xrightarrow{fission} {}^{47}_{22}Ti$ $\xrightarrow{fission} {}^{52}_{22}Ti$		${}^{47}_{22}Ti$ ${}^{52}_{22}Ti$	$\}$ -0.007619
${}^{107}_{47}Ag - (3){}^{4}_{2}He \uparrow \xrightarrow{\alpha}$	${}^{95}_{41}Nb$	${}^{95}_{42}Mo$	**+0.008554**
${}^{107}_{47}Ag - (5){}^{4}_{2}He \uparrow \xrightarrow{\alpha}$	${}^{87}_{37}Ru$	${}^{87}_{38}Sr$	**+0.008554**
${}^{109}_{47}Ag - (3){}^{4}_{2}He \uparrow \xrightarrow{\alpha}$ ${}^{97}_{42}Mo \xrightarrow{fission} {}^{45}_{21}Sc$ $\xrightarrow{fission} {}^{52}_{21}Sc$	${}^{97}_{41}Nb$	${}^{97}_{42}Mo$ **(note2)** ${}^{45}_{21}Sc$ ${}^{52}_{24}Cr$	**+0.009079** -6.985118 **H** +6.984072 **H**
${}^{109}_{47}Ag - (4){}^{4}_{2}He \uparrow \xrightarrow{\alpha}$	${}^{93}_{38}Y$	${}^{93}_{41}Nb$	**+0.012039**
${}^{64}_{28}Ni - (2){}^{4}_{2}He \xrightarrow{\alpha}$	${}^{56}_{24}Cr$	${}^{56}_{26}Fe$	**+0.0121778**
${}^{107}_{47}Ag \xrightarrow{fission}$	${}^{45}_{21}Sc$ ${}^{62}_{26}Fe$	${}^{62}_{28}Ni$	$\}$ **-0.01241**
${}^{58}_{28}Ni + {}^{64}_{28}Ni - (2){}^{4}_{2}He \xrightarrow{fusion}$ ${}^{114}_{50}Sn \xrightarrow{fission}$	${}^{114}_{52}Te$	$\left[{}^{114}_{50}Sn\right]$ $\left[(2){}^{57}_{26}Fe\right]$	+0.044676 **+0.012685**
$(2){}^{60}_{28}Ni + {}^{2}_{1}H^{+} \xrightarrow{fusion}$ ${}^{122}_{52}Te \xrightarrow{fision}$	${}^{122}_{57}La$	${}^{118}_{50}Sn, {}^{121}_{51}Sb, \left[{}^{122}_{52}Te\right]$ **absent (note 2)** $(2){}^{61}_{28}Ni$	+0.027369 **+0.013562**
${}^{68}_{30}Zn + (1){}^{2}_{1}H \xrightarrow{fusion}$	${}^{70}_{31}Ga$	${}^{70}_{30}Zn$	**-0.013626**
${}^{61}_{28}Ni + {}^{107}_{47}Ag + (2){}^{2}_{1}H^{+} \xrightarrow{fusion}$ ${}^{114}_{50}Sn - {}^{4}_{2}He \xrightarrow{\alpha}$ ${}^{112}_{50}Sn \xrightarrow{\beta^{+}\beta^{+}}$ ${}^{113}_{49}In - {}^{4}_{2}He \xrightarrow{\alpha}$ ${}^{114}_{50}Sn \xrightarrow{fission}$ **(note 2)**	${}^{114}_{55}Cs$	${}^{106}_{48}Cd, {}^{107}_{47}Ag, {}^{108}_{48}Cd, {}^{109}_{47}Ag, {}^{110}_{48}Cd,$ ${}^{111}_{48}Cd, {}^{112}_{50}Sn, {}^{113}_{49}In, {}^{114}_{50}Sn$ ${}^{110}_{48}Cd$ ${}^{112}_{48}Cd$ ${}^{109}_{47}Ag$ $\left[(2){}^{57}_{26}Fe\right]$	 -3.95358 -1.951766 -4.951832 **-0.014203**
	${}^{107}_{48}Cd$	${}^{109}_{47}Ag$	**-0.014446**

${}^{58}_{28}Ni-(2)^{4}_{2}He\xrightarrow{\alpha}$	${}^{50}_{24}Cr\xrightarrow{\beta^{+}\beta^{+}}$ **(note 5)**	${}^{50}_{22}Ti$	**+0.014654**
${}^{109}_{47}Ag+(1)^{2}_{1}H\xrightarrow{fusion}$	${}^{111}_{48}Cd$		**-0.014675**
${}^{65}_{29}Cu+(2)^{2}_{1}H\xrightarrow{fusion}$	${}^{67}_{30}Zn$	${}^{67}_{30}Zn$	**-0.014763**
${}^{61}_{28}Ni-(2)^{4}_{2}He\xrightarrow{\alpha}$	${}^{53}_{24}Cr$		**+0.0147999**
${}^{59}_{28}Ni-(2)^{4}_{2}He\xrightarrow{\alpha}$	${}^{51}_{24}Cr$	${}^{51}_{23}V$	**+0.0148193**
${}^{60}_{28}Ni-(3)^{4}_{2}He\xrightarrow{\alpha}$ ${}^{48}_{22}Ti+{}^{2}_{1}H^{+}\xrightarrow{fusion}$ ${}^{50}_{23}V\xrightarrow{\beta^{+}}$ $\xrightarrow{\beta^{-}}$	${}^{48}_{22}Ti$ **abse nt**	${}^{50}_{23}V$ ${}^{50}_{22}Ti$ ${}^{50}_{24}Cr$	+0.024969 **-0.014889** -0.0172568 -0.0160038
${}^{60}_{28}Ni-(2)^{4}_{2}He\xrightarrow{\alpha}$	${}^{52}_{24}Cr$		**+0.0149276**
${}^{60}_{28}Ni+{}^{61}_{28}Ni-{}^{4}_{2}He\xrightarrow{fusion}$	${}^{117}_{54}Xe$	${}^{116}_{50}Sn,\left[{}^{117}_{50}Sn\right]$	**+0.015139**
${}^{58}_{28}Ni+{}^{60}_{28}Ni-(2)^{4}_{2}He\xrightarrow{fusion}$ ${}^{110}_{48}Cd\xrightarrow{fission}$	${}^{110}_{52}Te$	${}^{109}_{47}Ag,\left[{}^{110}_{48}Cd\right]$ **absent** $\left[(2)^{55}_{24}Mn\right]$	+0.042079 **+0.015167**
${}^{66}_{30}Zn+(1)^{2}_{1}H\xrightarrow{fusion}$	${}^{68}_{31}Zn$	${}^{68}_{31}Zn$	**-0.015290**
${}^{61}_{28}Ni+(1)^{2}_{1}H\xrightarrow{fusion}$	${}^{63}_{29}Cu$	${}^{63}_{29}Cu$	**-0.015560**
${}^{67}_{30}Zn+(1)^{2}_{1}H\xrightarrow{fusion}$	${}^{69}_{31}Ga$	${}^{69}_{31}Ga$	**-0.015655**
${}^{109}_{47}Ag-(5)^{4}_{2}He\uparrow\xrightarrow{\alpha}$	${}^{89}_{37}Ru$	${}^{89}_{38}Y$	**+0.015714**
${}^{63}_{29}Cu+(2)^{2}_{1}H\xrightarrow{fusion}$	${}^{65}_{30}Zn$	${}^{65}_{29}Cu$	**-0.015909**
${}^{64}_{28}Ni+(1)^{2}_{1}H\xrightarrow{fusion}$	${}^{66}_{29}Cu$	${}^{66}_{30}Zn$	**-0.016034**
${}^{64}_{28}Ni\xrightarrow{fission}$	$(2)^{32}_{14}Si$	$(2)^{32}_{16}S$	**+0.016176**
${}^{59}_{27}Co+(1)^{2}_{1}H\xrightarrow{fusion}$	${}^{61}_{28}Ni$		**-0.0162405**

${}^{66}_{30}Zn+(2){}^{2}_{1}H\xrightarrow{fusion}$ ${}^{70}_{32}Ge\xrightarrow{fission}$	${}^{70}_{32}Ge$ **absent**	$(2){}^{35}_{17}Cl\uparrow$	-0.029969 **-0.016531**
${}^{60}_{28}Ni+(1){}^{2}_{1}H\xrightarrow{fusion}$	${}^{62}_{29}Cu$	${}^{62}_{28}Ni$	**-0.016543**
${}^{60}_{28}Ni\xrightarrow{fission}$	$(2){}^{30}_{14}Si$		**+0.0167539**
$(2){}^{58}_{28}Ni+{}^{2}_{1}H^{+}\xrightarrow{fusion}$	${}^{118}_{57}La$	${}^{114}_{50}Sn,{}^{116}_{50}Sn,{}^{117}_{50}Sn,\left[{}^{118}_{50}Sn\right]$	**+0.016815**
${}^{64}_{30}Zn+(1){}^{2}_{1}H\xrightarrow{fusion}$	${}^{66}_{31}Ga$	${}^{66}_{30}Zn$	**-0.017210**
${}^{70}_{30}Zn+(1){}^{2}_{1}H\xrightarrow{fusion}$	${}^{72}_{31}Ga$	${}^{72}_{32}Ge$	**-0.017345**
${}^{58}_{28}Ni\xrightarrow{fission}$	$(2){}^{29}_{14}Si$		**+0.0176465**
${}^{58}_{28}Ni+(1){}^{2}_{1}H\xrightarrow{fusion}$	${}^{60}_{29}Cu$	${}^{60}_{28}Ni$	**-0.018658**
${}^{58}_{28}Ni+{}^{62}_{28}Ni-(2){}^{4}_{2}He\xrightarrow{fusion}$ ${}^{112}_{50}Sn\xrightarrow{fission}$	${}^{112}_{52}Te$	$\left[{}^{112}_{50}Sn\right]$ $\left[(2){}^{56}_{26}Fe\right]$	+0.046336 **+0.019328**
${}^{64}_{28}Ni-(3){}^{4}_{2}He\xrightarrow{\alpha}$	${}^{52}_{22}Ti$	${}^{52}_{24}Cr$	**+0.0203510**
${}^{109}_{47}Ag-(6){}^{4}_{2}He\uparrow\xrightarrow{\alpha}$	${}^{85}_{35}Br\uparrow^{2.9\,min}$	${}^{85}_{37}Rb$	**+0.022657**
${}^{62}_{28}Ni-(3){}^{4}_{2}He\xrightarrow{\alpha}$	${}^{50}_{22}Ti$		**+0.0242558**
${}^{61}_{28}Ni-(3){}^{4}_{2}He\xrightarrow{\alpha}$	${}^{49}_{22}Ti$	${}^{49}_{22}Ti$	**+0.0246237**
${}^{107}_{47}Ag\xrightarrow{fission}$	${}^{53}_{22}Ti$ ${}^{54}_{25}Mn$	${}^{53}_{24}Cr$ $\xrightarrow{EC(99.99\%)}{}^{54}_{24}Cr$ $\xrightarrow{\beta^-(3x10^{-4}\%)}{}^{54}_{26}Fe$ $\xrightarrow{\beta^+(6x10^{-7}\%)}{}^{54}_{24}Cr$	-0.0255672 **-0.0248371** -0.0255672
${}^{58}_{28}Ni-(3){}^{4}_{2}He\xrightarrow{\alpha}$	${}^{46}_{22}Ti$	${}^{46}_{22}Ti$	**+0.0250984**
${}^{59}_{28}Ni-(3){}^{4}_{2}He\xrightarrow{\alpha}$	${}^{47}_{22}Ti$	${}^{47}_{22}Ti$	**+0.0252261**

$^{107}_{47}Ag \xrightarrow{fission}$	$^{53}_{23}V$ $^{54}_{24}Cr$	$^{53}_{24}Cr$	} **-0.0255672**
$^{107}_{47}Ag \xrightarrow{fission}$	$^{54}_{23}V$ $^{53}_{24}Cr$	$^{54}_{24}Cr$	} **-0.0255672**
$^{107}_{47}Ag \xrightarrow{fission}$	$^{54}_{22}Ti$ $^{53}_{25}Mn$	$^{54}_{24}Cr$ $^{53}_{24}Cr$	} **-0.0255672**
$^{107}_{47}Ag - (6)^{4}_{2}He \uparrow \xrightarrow{\alpha}$	$^{83}_{35}Br \uparrow^{2.9\,min}$	$^{85}_{37}Rb$	**+0.025632**
$^{107}_{47}Ag \xrightarrow{fission}$	$^{51}_{23}V$ $^{56}_{24}Cr$	$^{56}_{26}Fe$	} **-0.0262000**
$^{107}_{47}Ag \xrightarrow{fission}$	$^{51}_{22}Ti$ $^{56}_{25}Mn$	$^{51}_{23}V$ $^{56}_{26}Fe$	} **-0.0262000**
$^{107}_{47}Ag \xrightarrow{fission}$	$^{52}_{23}V$ $^{55}_{24}Cr$	$^{52}_{24}Cr$ $^{55}_{25}Mn$	} **-0.0265438**
$^{107}_{47}Ag \xrightarrow{fission}$	$^{55}_{23}V$ $^{52}_{24}Cr$	$^{55}_{25}Mn$	} **-0.0265438**
$^{107}_{47}Ag \xrightarrow{fission}$	$^{52}_{22}Ti$ $^{55}_{25}Mn$	$^{52}_{24}Cr$	} **-0.0265438**
$^{107}_{47}Ag \xrightarrow{fission}$	$^{55}_{22}Ti$ $^{52}_{25}Mn$	$^{55}_{25}Mn$ $^{52}_{24}Cr$	} **-0.0265438**
$^{109}_{47}Ag \xrightarrow{fission}$	$^{50}_{20}Ca$ $^{59}_{27}Co$	$^{50}_{22}Ti$	} **-0.0267658**
$^{109}_{47}Ag \xrightarrow{fission}$	$^{55}_{22}Ti$ $^{54}_{25}Mn$	$^{55}_{25}Mn$ $\xrightarrow{EC(99.99\%)} {}^{54}_{24}Cr$ $\xrightarrow{\beta^{-}(3x10^{-4}\%)} {}^{54}_{26}Fe$	 +0.02782582 **+0.0270965**

		$\xrightarrow{\beta^{+}(6x10^{-7}\%)} {}^{54}_{24}Cr$	+0.02782582
${}^{62}_{28}Ni - (1)^{4}_{2}He \xrightarrow{\alpha}$	${}^{58}_{26}Fe \xrightarrow{fission}$	$(2)^{29}_{14}Si$	**+0.0272476**
${}^{109}_{47}Ag + (2)^{2}_{1}H \xrightarrow{fusion}$	${}^{113}_{49}In$		**-0.028897**
${}^{107}_{47}Ag + (2)^{2}_{1}H \xrightarrow{fusion}$	${}^{111}_{49}In$	${}^{111}_{48}Cd$	**-0.029121**
${}^{109}_{47}Ag \xrightarrow{fission}$	${}^{53}_{22}Ti$ ${}^{56}_{25}Mn$	${}^{53}_{24}Cr$ ${}^{56}_{26}Fe$	} **-0.0291651**
${}^{109}_{47}Ag \xrightarrow{fission}$	${}^{56}_{22}Ti$ ${}^{53}_{25}Mn$	${}^{56}_{26}Fe$ ${}^{53}_{24}Cr$	} **-0.0291651**
${}^{109}_{47}Ag \xrightarrow{fission}$	${}^{53}_{23}V$ ${}^{56}_{24}Cr$	${}^{53}_{24}Cr$ ${}^{56}_{26}Fe$	} **-0.0291651**
${}^{109}_{47}Ag \xrightarrow{fission}$	${}^{56}_{23}V$ ${}^{53}_{24}Cr$	${}^{56}_{26}Fe$	} **-0.0291651**
${}^{107}_{47}Ag - (7)^{4}_{2}He \uparrow \xrightarrow{\alpha}$	${}^{79}_{33}As$	${}^{79}_{35}Br \uparrow$	**+0.029649**
${}^{109}_{47}Ag - (7)^{4}_{2}He \uparrow \xrightarrow{\alpha}$	${}^{81}_{33}As$	${}^{81}_{35}Br \uparrow$	**+0.029761**
${}^{109}_{47}Ag \xrightarrow{fission}$	${}^{54}_{22}Ti$ ${}^{55}_{25}Mn$	${}^{54}_{24}Cr$	} **-0.0298266**
${}^{60}_{28}Ni + (2)^{2}_{1}H \xrightarrow{fusion}$	${}^{64}_{30}Zn$	${}^{64}_{30}Zn$	**-0.029847**
${}^{65}_{29}Cu + (3)^{2}_{1}H \xrightarrow{fusion}$	${}^{69}_{31}Ga$	${}^{69}_{31}Ga$	**-0.030418**
${}^{62}_{28}Ni + (2)^{2}_{1}H \xrightarrow{fusion}$	${}^{66}_{30}Zn$	${}^{66}_{30}Zn$	**-0.030515**
${}^{67}_{30}Zn + (2)^{2}_{1}H \xrightarrow{fusion}$	${}^{71}_{32}Ge$	${}^{71}_{31}Ga$	**-0.030629**
${}^{63}_{29}Cu + (3)^{2}_{1}H \xrightarrow{fusion}$	${}^{67}_{31}Ga$	${}^{67}_{30}Zn$	**-0.030673**
${}^{68}_{30}Zn + (2)^{2}_{1}H \xrightarrow{fusion}$	${}^{72}_{32}Ge$	${}^{72}_{32}Ge$	**-0.030971**
${}^{64}_{28}Ni + (2)^{2}_{1}H \xrightarrow{fusion}$	${}^{68}_{30}Zn$	${}^{68}_{30}Zn$	**-0.031325**

${}^{61}_{28}Ni + (2)^2_1H \xrightarrow{fusion}$	${}^{65}_{30}Zn$	${}^{65}_{29}Cu$	**-0.031470**
${}^{59}_{27}Co + (2)^2_1H \xrightarrow{fusion}$	${}^{63}_{29}Cu$		**-0.0318005**
${}^{70}_{30}Zn + (2)^2_1H \xrightarrow{fusion}$	${}^{74}_{32}Ge$	${}^{74}_{32}Ge$	**-0.032344**
${}^{64}_{30}Zn + (2)^2_1H \xrightarrow{fusion}$	${}^{68}_{32}Ge$	${}^{68}_{30}Zn$	**-0.032501**
${}^{64}_{28}Ni + (3)^2_1H \xrightarrow{fusion}$ ${}^{70}_{31}Ge \xrightarrow{fission}$	${}^{70}_{31}Ga$	$\xrightarrow{EC(.4\%)} {}^{70}_{30}Zn$ $\xrightarrow{\beta^-(99.6\%)} {}^{70}_{32}Ge$ **absent** $(2)^{35}_{17}Cl \uparrow$	-0.044952 -0.046022 **-0.032565**
$(3)^{64}_{28}Ni \xrightarrow{fusion}$ ${}^{192}_{84}Pt \xrightarrow{fission}$	${}^{192}_{84}Po$ $(2)^{96}_{39}Y$ $(2)^{97}_{39}Y$	${}^{176}_{72}Hf, {}^{184}_{76}Os, {}^{188}_{76}Os, [{}^{192}_{78}Pt]$ $[(2)^{96}_{40}Zr]$ $[(2)^{97}_{42}Mo]$	+0.207437 **+0.032648** +2.028145
${}^{64}_{30}Zn + (3)^2_1H \xrightarrow{fusion}$ ${}^{70}_{32}Ge \xrightarrow{fission}$	${}^{70}_{33}As$	${}^{70}_{32}Ge$ **absent** $(2)^{35}_{17}Cl \uparrow$	-0.047200 **-0.033743**
${}^{58}_{28}Ni + (2)^2_1H \xrightarrow{fusion}$	${}^{62}_{30}Zn$	${}^{62}_{28}Ni$	**-0.035201**
${}^{68}_{30}Zn + (3)^2_1H \xrightarrow{fusion}$ ${}^{74}_{34}Se \xrightarrow{fission}$	${}^{74}_{33}As$	$\xrightarrow{\beta^+(66\%)} {}^{74}_{32}Ge$ $\xrightarrow{\beta^-(34\%)} {}^{74}_{34}Se$ $(2)^{37}_{17}Cl \uparrow$	-0.045971 -0.044672 **-0.035344**
${}^{109}_{47}Ag - (8)^4_2He \uparrow \xrightarrow{\alpha}$	${}^{75}_{31}Ga$	${}^{77}_{34}Se$	**+0,035988**
${}^{107}_{47}Ag - (8)^4_2He \uparrow \xrightarrow{\alpha}$	${}^{71}_{29}Cu$	${}^{75}_{33}As$	**+0.037325**
${}^{58}_{28}Ni + {}^{61}_{28}Ni - (2)^4_2He \xrightarrow{fusion}$	${}^{111}_{52}Te$	${}^{110}_{48}Cd, [{}^{111}_{48}Cd]$	**+0.042985**
${}^{107}_{47}Ag + (3)^2_1H \xrightarrow{fusion}$	${}^{113}_{50}Sn$	${}^{113}_{49}In$	**-0.043343**
${}^{60}_{28}Ni + {}^{62}_{28}Ni - {}^4_2He \xrightarrow{fusion}$	${}^{118}_{54}Xe$	$[{}^{118}_{50}Sn]$	**+0.045074**

${}^{65}_{29}Cu+(4)^2_1H \xrightarrow{fusion}$	${}^{71}_{32}Ge$	${}^{71}_{31}Ga$	**-0.045392**
${}^{70}_{30}Zn+(4)^2_1H \xrightarrow{fusion}$ $(2)^{39}_{17}Cl \uparrow^{59\,min} \xrightarrow{\beta^-}$	${}^{78}_{34}Se \xrightarrow{fission}$ **absent**	 $(2)^{39}_{17}Cl \uparrow^{59\,min}$ $(2)^{39}_{18}Ar \uparrow^{269yr}$	-0.064417 **-0.045710** -0.053099
${}^{62}_{28}Ni+(3)^2_1H \xrightarrow{fusion}$	${}^{68}_{31}Ge$	${}^{68}_{30}Zn$	**-0.045806**
${}^{67}_{30}Zn+(3)^2_1H \xrightarrow{fusion}$	${}^{73}_{33}As$	${}^{73}_{32}Ge$	**-0.045973**
${}^{70}_{30}Zn+(3)^2_1H \xrightarrow{fusion}$ ${}^{76}_{32}Ge \xrightarrow{\beta^-\beta^-}$	${}^{76}_{33}As$	$\xrightarrow{\beta^-(99+\%)} {}^{76}_{34}Se$ $\xrightarrow{EC(.02\%)} {}^{76}_{32}Ge$ ${}^{76}_{34}Se$	-0.048411 **-0.046222** -0.048411
${}^{61}_{28}Ni+(3)^2_1H \xrightarrow{fusion}$	${}^{67}_{31}Ga$	${}^{67}_{30}Zn$	**-0.046234**
${}^{66}_{30}Zn+(3)^2_1H \xrightarrow{fusion}$	${}^{72}_{33}As$	${}^{72}_{32}Ge$	**-0.046262**
${}^{63}_{29}Cu+(4)^2_1H \xrightarrow{fusion}$	${}^{69}_{32}Ge$	${}^{69}_{31}Ga$	**-0.046328**
${}^{62}_{28}Ni+(4)^2_1H \xrightarrow{fusion}$	${}^{70}_{32}Ge \xrightarrow{fission}$	${}^{35}_{17}Cl \uparrow$	**-0.047048**
${}^{60}_{28}Ni+(3)^2_1H \xrightarrow{fusion}$	${}^{66}_{31}Ga$	${}^{66}_{30}Zn$	**-0.047058**
$(2)^{64}_{28}Ni \xrightarrow{fusion}$	${}^{128}_{56}Ba$	$\left[{}^{128}_{54}Xe\right]$	**-0.0475993**
${}^{59}_{27}Co+(3)^2_1H \xrightarrow{fusion}$	${}^{65}_{30}Zn$	${}^{65}_{29}Cu$	**-0.0477101**
${}^{58}_{28}Ni+(3)^2_1H \xrightarrow{fusion}$	${}^{64}_{31}Ga$	${}^{64}_{30}Zn$	**-0.048506**
$(2)^{62}_{28}Ni \xrightarrow{fusion}$	${}^{124}_{56}Ba$	$\left[{}^{124}_{54}Xe\right]$	**+0.0492028**
${}^{66}_{30}Zn+(4)^2_1H \xrightarrow{fusion}$ ${}^{74}_{34}Se \xrightarrow{fission}$	${}^{74}_{34}Se$	$(2)^{37}_{17}Cl$	-0.059963 **-0.050634**
${}^{107}_{47}Ag+(4)^2_1H \xrightarrow{fusion}$ ${}^{115}_{51}Sb \xrightarrow{\beta^+} {}^{115}_{50}Sn$ **absent(note1)**	${}^{115}_{51}Sb$	${}^{111}_{49}Cd$	-0.0549061 **-0.0547226**

${}^{115}_{51}Sb - {}^{4}_{2}He \xrightarrow{\alpha}$			
${}^{109}_{47}Ag + (4){}^{2}_{1}H \xrightarrow{fusion}$	${}^{117}_{51}Sb$	${}^{117}_{50}Sn$	**-0.0582061**
${}^{66}_{30}Zn + (5){}^{2}_{1}H \xrightarrow{fusion}$ ${}^{76}_{34}Se \xrightarrow{fission}$ $(2){}^{38}_{17}Cl \uparrow^{37\,min} \xrightarrow{\beta^-}$	${}^{76}_{35}Br \uparrow^{16hrs}$	${}^{76}_{34}Se$ $(2){}^{38}_{17}Cl \uparrow^{37\,min}$ $(2){}^{40}_{17}Ar \uparrow$	-0.077328 **-0.060521** -0.071077
${}^{65}_{29}Cu + (5){}^{2}_{1}H \xrightarrow{fusion}$	${}^{73}_{33}As$	${}^{73}_{32}Ge$	**-0.060736**
${}^{63}_{29}Cu + (5){}^{2}_{1}H \xrightarrow{fusion}$	${}^{71}_{33}As$	${}^{71}_{31}Ga$	**-0.061302**
${}^{67}_{30}Zn + (4){}^{2}_{1}H \xrightarrow{fusion}$	${}^{75}_{34}Se$	${}^{75}_{33}As$	**-0.061636**
${}^{61}_{28}Ni + (4){}^{2}_{1}H \xrightarrow{fusion}$	${}^{69}_{32}Ge$	${}^{69}_{31}Ga$	**-0.061889**
${}^{68}_{30}Zn + (4){}^{2}_{1}H \xrightarrow{fusion}$	${}^{76}_{34}Se$		**-0.062036**
${}^{64}_{28}Ni + (4){}^{2}_{1}H \xrightarrow{fusion}$	${}^{72}_{32}Ge$		**-0.062297**
${}^{60}_{28}Ni + (4){}^{2}_{1}H \xrightarrow{fusion}$	${}^{68}_{32}Ge$	${}^{68}_{30}Zn$	**-0.062349**
${}^{59}_{27}Co + (4){}^{2}_{1}H \xrightarrow{fusion}$	${}^{67}_{31}Ga$	${}^{67}_{30}Zn$	**-0.0624738**
${}^{64}_{30}Zn + (4){}^{2}_{1}H \xrightarrow{fusion}$	${}^{72}_{34}Se$	${}^{72}_{32}Ge$	**-0.063472**
${}^{60}_{28}Ni + (5){}^{2}_{1}H \xrightarrow{fusion}$ ${}^{70}_{32}Ge \xrightarrow{fission}$	${}^{70}_{33}As$	${}^{70}_{32}Ge$ $(2){}^{35}_{16}S \xrightarrow{\beta -} (2){}^{35}_{17}Cl \uparrow$	-0.377049 **-0.063579**
${}^{58}_{28}Ni + (4){}^{2}_{1}H \xrightarrow{fusion}$	${}^{66}_{32}Ge$	${}^{66}_{30}Zn$	**-0.065716**
${}^{64}_{28}Ni + (5){}^{2}_{1}H \xrightarrow{fusion}$ ${}^{74}_{34}Se \xrightarrow{fission}$	${}^{74}_{33}As \xrightarrow{17day}$	$\xrightarrow{\beta^+(66\%)} {}^{74}_{32}Ge$ $\xrightarrow{\beta^-(34\%)} {}^{74}_{34}Se$ **absent** $(2){}^{37}_{17}Cl$	-0.077293 -0.075994 **-0.066666**
${}^{64}_{30}Zn + (5){}^{2}_{1}H \xrightarrow{fusion}$ ${}^{74}_{34}Se \xrightarrow{fission}$	${}^{74}_{35}Br \uparrow^{25\,min}$	${}^{74}_{34}Se$ **absent** $(2){}^{37}_{17}Cl \uparrow$	-0.077174 **-0.067825**
${}^{109}_{47}Ag + (5){}^{2}_{1}H \xrightarrow{fusion}$	${}^{119}_{52}Te$	${}^{119}_{50}Sn$	**-0.0719517**
${}^{61}_{28}Ni + {}^{107}_{47}Ag + (2){}^{2}_{1}H^{+} \xrightarrow{fusion}$	${}^{172}_{77}Ir$	${}^{164}_{68}Er, {}^{168}_{70}Yb, \left[{}^{172}_{70}Yb\right]$	**+0.0720249**

${}^{107}_{47}Ag + (5)^2_1H \xrightarrow{fusion}$	${}^{117}_{52}Te$	${}^{117}_{50}Sn$	**-0.0726527**
${}^{68}_{30}Zn + (5)^2_1H \xrightarrow{fusion}$	${}^{78}_{35}Br \uparrow^{6\,min}$	$\xrightarrow{\beta^+(99+\%)} {}^{78}_{34}Se$ $\xrightarrow{\beta^-(.01\%)} {}^{78}_{36}Kr \uparrow$	-0.078043 **-0.074988**
${}^{63}_{29}Cu + (6)^2_1H \xrightarrow{fusion}$	${}^{73}_{34}Se$	${}^{73}_{32}Ge$	**-0.076646**
${}^{65}_{29}Cu + (6)^2_1H \xrightarrow{fusion}$	${}^{75}_{34}Se$	${}^{75}_{33}As$	**-0.076700**
${}^{62}_{28}Ni + (5)^2_1H \xrightarrow{fusion}$	${}^{72}_{33}As$	${}^{72}_{31}Ge$	**-0.076778**
${}^{58}_{28}Ni + {}^{107}_{47}Ag + {}^2_1H^+ \xrightarrow{fusion}$ ${}^{167}_{68}Er \xrightarrow{\alpha}$	${}^{167}_{76}Os$	${}^{143}_{60}Nd, {}^{147}_{62}Sm, {}^{155}_{64}Gd, {}^{159}_{65}Tb, {}^{163}_{66}Dy, \left[{}^{167}_{68}Er\right]$ $\left[{}^{163}_{66}Dy\right]$	+0.0775065 **+0.0767937**
${}^{61}_{28}Ni + (5)^2_1H \xrightarrow{fusion}$	${}^{71}_{33}As$	${}^{71}_{33}Ga$	**-0.076863**
${}^{67}_{30}Zn + (5)^2_1H \xrightarrow{fusion}$	${}^{77}_{35}Br \uparrow^{57hrs}$	${}^{77}_{34}Se$	**-0.077721**
${}^{59}_{27}Co + (5)^2_1H \xrightarrow{fusion}$	${}^{69}_{32}Ge$	${}^{69}_{31}Ga$	**-0.0781291**
${}^{109}_{47}Ag + (3)^2_1H \xrightarrow{fusion}$ ${}^{115}_{50}Sn \xrightarrow{fission} {}^{56}_{25}Mn$ and	${}^{115}_{50}Sn$ **note** **2**	${}^{56}_{26}Fe$ ${}^{59}_{27}Co$	} **-0.078924**
${}^{70}_{30}Zn + (5)^2_1H \xrightarrow{fusion}$ •	${}^{80}_{35}Br \uparrow^{18\,min}$	$\xrightarrow{\beta^-(92\%)} {}^{80}_{36}Kr \uparrow$ $\xrightarrow{\beta^+(8\%)} {}^{80}_{34}Se$	-0.079449 **-0.079306**
${}^{58}_{28}Ni + (5)^2_1H \xrightarrow{fusion}$	${}^{68}_{33}As$	${}^{68}_{30}Zn$	**-0.081007**
${}^{62}_{28}Ni + (6)^2_1H \xrightarrow{fusion}$	${}^{74}_{34}Se \xrightarrow{fission}$	${}^{37}_{17}Cl \uparrow$	**-0.081150**
${}^{58}_{28}Ni + (6)^2_1H \xrightarrow{fusion}$ ${}^{70}_{32}Ge \xrightarrow{fission}$	${}^{70}_{34}Se$	${}^{70}_{32}Ge$ **absent** $(2)^{35}_{16}S \xrightarrow{\beta-} (2)^{35}_{17}Cl \uparrow$	-0.095706 **-0.082248**
${}^{107}_{47}Ag + (6)^2_1H \xrightarrow{fusion}$	${}^{119}_{53}I \uparrow^{19\,min}$	${}^{119}_{50}Sn$	**-0.0838612**
${}^{109}_{47}Ag + (6)^2_1H \xrightarrow{fusion}$	${}^{121}_{53}I \uparrow^{2hrs}$	${}^{121}_{51}Sb$	**-0.0855455**

${}^{66}_{30}Zn + (6)^{2}_{1}H \xrightarrow{fusion}$	${}^{78}_{36}Kr \uparrow$	${}^{78}_{36}Kr \uparrow$	**-0.090277**
${}^{64}_{28}Ni + (7)^{2}_{1}H \xrightarrow{fusion}$ ${}^{78}_{34}Se \xrightarrow{fission}$	${}^{78}_{35}Br \uparrow^{6\,min}$	$\xrightarrow{\beta^{+}(99.9\%)} {}^{78}_{34}Se$ **absent** $\xrightarrow{\beta^{-}(0.01\%)} {}^{78}_{36}Kr \uparrow$ ${}^{39}_{18}Ar \uparrow^{2093\,yr}$	-0.109369 -0.106313 **+0.090805**
${}^{61}_{28}Ni + (6)^{2}_{1}H \xrightarrow{fusion}$	${}^{73}_{34}Se$	${}^{73}_{32}Ge$	**-0.092207**
${}^{65}_{29}Cu + (7)^{2}_{1}H \xrightarrow{fusion}$	${}^{77}_{35}Br \uparrow^{57\,hrs}$	${}^{77}_{34}Se$	**-0.092484**
${}^{63}_{29}Cu + (7)^{2}_{1}H \xrightarrow{fusion}$	${}^{75}_{35}Br \uparrow^{96\,min}$	${}^{75}_{33}As$	**-0.092610**
${}^{68}_{30}Zn + (6)^{2}_{1}H \xrightarrow{fusion}$	${}^{80}_{36}Kr \uparrow$	${}^{80}_{36}Kr \uparrow$	**-0.093074**
${}^{59}_{27}Co + (6)^{2}_{1}H \xrightarrow{fusion}$	${}^{71}_{33}As$	${}^{71}_{31}Ga$	**-0.0931029**
${}^{60}_{28}Ni + (6)^{2}_{1}H \xrightarrow{fusion}$	${}^{72}_{34}Se$	${}^{72}_{32}Ge$	**-0.093321**
${}^{64}_{28}Ni + (6)^{2}_{1}H \xrightarrow{fusion}$	${}^{76}_{34}Se$		**-0.093363**
${}^{67}_{30}Zn + (6)^{2}_{1}H \xrightarrow{fusion}$	${}^{79}_{36}Kr \uparrow^{35\,hrs}$	${}^{79}_{35}Br \uparrow$	**-0.093399**
${}^{64}_{30}Zn + (6)^{2}_{1}H \xrightarrow{fusion}$	${}^{76}_{36}Kr \uparrow^{14.8\,hrs}$	${}^{76}_{34}Se$	**-0.094537**
${}^{70}_{30}Zn + (6)^{2}_{1}H \xrightarrow{fusion}$	${}^{82}_{36}Kr \uparrow$		**-0.096444**
${}^{109}_{47}Ag + (7)^{2}_{1}H \xrightarrow{fusion}$ ${}^{123}_{52}Te \xrightarrow{\beta^{+}}$ ${}^{123}_{52}Te - {}^{4}_{2}He \xrightarrow{\alpha}$	${}^{123}_{54}Xe \uparrow^{2.1\,hr}$	${}^{123}_{52}Te$ **absent(note 4)** ${}^{123}_{51}Sb$ **absent** ${}^{119}_{50}Sn$	**-0.0991943** **-0.0992503** **-0.0975531**
${}^{60}_{28}Ni + (7)^{2}_{1}H \xrightarrow{fusion}$ ${}^{74}_{34}Se \xrightarrow{fission}$	${}^{74}_{35}Br \uparrow^{25\,min}$	${}^{74}_{34}Se$ **absent** $(2)^{37}_{17}Cl \uparrow$	-0.107011 **-0.097682**
${}^{107}_{47}Ag + (7)^{2}_{1}H \xrightarrow{fusion}$	${}^{121}_{54}Xe \uparrow^{40\,min}$	${}^{121}_{51}Sb$	**-0.0999920**

${}^{70}_{30}Zn+(7)^{2}_{1}H\xrightarrow{fusion}$ ${}^{64}_{38}Sr\xrightarrow{fission}$ ${}^{64}_{38}Sr\xrightarrow{fission}$ ${}^{45}_{21}Sc+{}^{39}_{17}Cl\uparrow^{56\,min}\rightarrow$	${}^{84}_{37}Rb$	$\xrightarrow{\beta^{+}(96\%)}{}^{84}_{36}Kr\uparrow$ $\xrightarrow{\beta^{-}(4\%)}{}^{84}_{38}Sr$ **absent** $(2)^{42}_{20}Ca$ **absent** ${}^{45}_{21}Sc+{}^{39}_{17}Cl\uparrow^{56\,min}$ ${}^{45}_{21}Sc+{}^{19}_{9}Fl\uparrow+{}^{20}_{10}Ne\uparrow$	-0.112524 -0.110606 -0.106795 **-0.100111** -0.577276
${}^{64}_{30}Zn+(7)^{2}_{1}H\xrightarrow{fusion}$	${}^{78}_{37}Rb$	${}^{78}_{36}Kr\uparrow$	**-0.107488**
${}^{62}_{28}Ni+(7)^{2}_{1}H\xrightarrow{fusion}$	${}^{76}_{35}Br\uparrow^{16hrs}$	${}^{76}_{34}Se$	**-0.107843**
	${}^{79}_{36}Kr\uparrow^{35hrs}$	${}^{79}_{35}Br\uparrow$	**-0.108163**
${}^{61}_{28}Ni+(7)^{2}_{1}H\xrightarrow{fusion}$	${}^{75}_{35}Br\uparrow^{96\,min}$	${}^{75}_{33}As$	**-0.108171**
${}^{66}_{30}Zn+(7)^{2}_{1}H\xrightarrow{fusion}$	${}^{80}_{37}Rb$	${}^{80}_{36}Kr\uparrow$	**-0.108365**
${}^{63}_{29}Cu+(8)^{2}_{1}H\xrightarrow{fusion}$	${}^{77}_{36}Kr\uparrow^{74\,min}$	${}^{77}_{34}Se$	**-0.108394**
${}^{59}_{27}Co+(7)^{2}_{1}H\xrightarrow{fusion}$	${}^{73}_{34}As$	${}^{73}_{32}Ge$	**-0.1084468**
${}^{67}_{30}Zn+(7)^{2}_{1}H\xrightarrow{fusion}$	${}^{81}_{37}Rb$	${}^{81}_{36}Kr\uparrow$	**-0.109246**
${}^{68}_{30}Zn+(7)^{2}_{1}H\xrightarrow{fusion}$	${}^{82}_{37}Kr\uparrow$	${}^{82}_{37}Kr\uparrow$	**-0.110071**
${}^{58}_{28}Ni+(7)^{2}_{1}H\xrightarrow{fusion}$	${}^{72}_{35}Br\uparrow^{79\,sec}$	${}^{72}_{32}Ge$	**-0.111979**
${}^{107}_{47}Ag+(8)^{2}_{1}H\xrightarrow{fusion}$ ${}^{123}_{52}Te\xrightarrow{\beta^{+}}$ ${}^{123}_{52}Te-{}^{4}_{2}He\xrightarrow{\alpha}$	${}^{123}_{55}Cs$	${}^{123}_{52}Te$ **absent(note 4)** ${}^{123}_{51}Sb$ **absent** ${}^{119}_{50}Sn$	-0.113640 -0.113697 **-0.111999**
${}^{109}_{49}Ag+(8)^{2}_{1}H\xrightarrow{fusion}$	${}^{125}_{55}Cs$	${}^{125}_{53}I\uparrow^{59days}\xrightarrow{EC}{}^{125}_{52}Te$	**-0.1131336**
${}^{58}_{28}Ni+(8)^{2}_{1}H\xrightarrow{fusion}$ ${}^{74}_{34}Se\xrightarrow{fission}$	${}^{74}_{36}Kr\uparrow^{11\,min}$	${}^{74}_{34}Se$ **absent** $(2)^{37}_{17}Cl\uparrow$	-0.125678 **-0.116351**

${}^{62}_{28}Ni + (8)^{2}_{1}H \xrightarrow{fusion}$	${}^{78}_{36}Kr \uparrow$	${}^{78}_{36}Kr \uparrow$	**-0.120794**
${}^{61}_{28}Ni + (8)^{2}_{1}H \xrightarrow{fusion}$	${}^{77}_{36}Kr \uparrow^{74\,min}$	${}^{77}_{33}Se$	**-0.123956**
${}^{63}_{29}Cu + (9)^{2}_{1}H \xrightarrow{fusion}$	${}^{79}_{37}Rb$	${}^{79}_{35}Br \uparrow$	**-0.124072**
${}^{60}_{28}Ni + (8)^{2}_{1}H \xrightarrow{fusion}$	${}^{76}_{36}Kr \uparrow^{76hr}$	${}^{76}_{34}Se$	**-0.124386**
${}^{64}_{28}Ni + (8)^{2}_{1}H \xrightarrow{fusion}$	${}^{80}_{36}Kr \uparrow^{stable}$	${}^{80}_{36}Kr \uparrow$	**-0.124401**
${}^{59}_{27}Co + (8)^{2}_{1}H \xrightarrow{fusion}$	${}^{75}_{35}Br \uparrow^{96\,min}$	${}^{75}_{33}As$	**-0.1244108**
${}^{66}_{30}Zn + (8)^{2}_{1}H \xrightarrow{fusion}$	${}^{82}_{38}Sr$	${}^{82}_{36}Kr \uparrow$	**-0.125362**
${}^{64}_{30}Zn + (8)^{2}_{1}H \xrightarrow{fusion}$	${}^{80}_{38}Sr$	${}^{80}_{36}Kr \uparrow$	**-0.125575**
${}^{67}_{30}Zn + (8)^{2}_{1}H \xrightarrow{fusion}$	${}^{83}_{38}Sr$	${}^{83}_{36}Kr \uparrow$	**-0.125803**
${}^{60}_{28}Ni + (9)^{2}_{1}H \xrightarrow{fusion}$	${}^{78}_{37}Rb$	${}^{78}_{36}Kr \uparrow$	**-0.137337**
${}^{62}_{28}Ni + (9)^{2}_{1}H \xrightarrow{fusion}$	${}^{80}_{37}Rb$	${}^{80}_{36}Kr \uparrow$	**-0.138882**
${}^{61}_{28}Ni + (9)^{2}_{1}H \xrightarrow{fusion}$	${}^{79}_{37}Rb$	${}^{79}_{35}Br \uparrow$	**-0.139634**
${}^{59}_{27}Co + (9)^{2}_{1}H \xrightarrow{fusion}$	${}^{77}_{36}Kr \uparrow^{4\,min}$	${}^{77}_{35}Br \uparrow^{57hr} \xrightarrow{\beta^{-}} {}^{77}_{34}Se$	**-0.1401948**
${}^{65}_{29}Cu + (1)^{2}_{1}H \xrightarrow{fusion}$	${}^{81}_{38}Sr$	${}^{81}_{35}Br \uparrow$	**-0.140220**
${}^{58}_{28}Ni + (9)^{2}_{1}H \xrightarrow{fusion}$	${}^{76}_{37}Rb$	${}^{76}_{34}Se$ ${}^{76}_{32}Ge$	-0.143045 **-0.140856**
${}^{64}_{28}Ni + (9)^{2}_{1}H \xrightarrow{fusion}$	${}^{80}_{37}Rb$	${}^{82}_{36}Kr \uparrow$	**-0.141398**
${}^{60}_{28}Ni + (10)^{2}_{1}H \xrightarrow{fusion}$	${}^{80}_{38}Sr$	${}^{80}_{36}Kr \uparrow$	**-0.155425**
${}^{61}_{28}Ni + (10)^{2}_{1}H \xrightarrow{fusion}$	${}^{81}_{32}Sr$	${}^{81}_{35}Br \uparrow$	**-0.155783**
${}^{62}_{28}Ni + (10)^{2}_{1}H \xrightarrow{fusion}$	${}^{84}_{38}Sr$	${}^{82}_{36}Kr$	**-0.155879**
${}^{58}_{28}Ni + (10)^{2}_{1}H \xrightarrow{fusion}$	${}^{78}_{38}Sr$	${}^{78}_{36}Kr \uparrow$	**-0.155995**

${}^{62}_{28}Ni - (2){}^{4}_{2}He \xrightarrow{\alpha}$	${}^{54}_{24}Cr$		**+0.1574181**
${}^{63}_{29}Cu + (1){}^{2}_{1}H \xrightarrow{fusion}$	${}^{84}_{38}Sr$	${}^{84}_{36}Kr \uparrow$	**-0.157476**
${}^{109}_{47}Ag \xrightarrow{fission}$	${}^{55}_{23}V$ ${}^{54}_{24}Cr$	${}^{55}_{25}Mn$	} **-0.9725126**

www.ingramcontent.com/pod-product-compliance
Lightning Source LLC
Chambersburg PA
CBHW050723180526
45159CB00003B/1114
* 9 7 8 1 4 9 1 0 8 1 4 8 8 *